The Fabric of Reality

The Fabric of Reality

*The Science of Parallel Universes—
and Its Implications*

DAVID DEUTSCH

ALLEN LANE
THE PENGUIN PRESS

ALLEN LANE
THE PENGUIN PRESS
Published by the Penguin Group
Penguin Putnam Inc., 375 Hudson Street,
New York, New York 10014, U.S.A.
Penguin Books Ltd, 27 Wrights Lane,
London W8 5TZ, England
Penguin Books Australia Ltd, Ringwood,
Victoria, Australia
Penguin Books Canada Ltd, 10 Alcorn Avenue,
Toronto, Ontario, Canada M4V 3B2
Penguin Books (N.Z.) Ltd, 182–190 Wairau Road,
Auckland 10, New Zealand

Penguin Books Ltd, Registered Offices:
Harmondsworth, Middlesex, England

First American edition
Published in 1997 by Allen Lane The Penguin Press,
an imprint of Viking Penguin, a member of Penguin Putnam Inc.

10 9 8 7 6 5 4 3 2 1

Figures drawn by Nigel Andrews

LIBRARY OF CONGRESS CATALOGING-IN-PUBLICATION DATA
Deutsch, David.
The fabric of reality / David Deutsch.
p. cm.
Includes bibliographical references and index.
ISBN 0-7139-9061-9 (alk. paper)
1. Reality. 2. Physics—Philosophy. 3. Life. 4. Cosmology. I. Title.
QC6.4.R42D48 1997
530′.01—dc21 97-6171

This book is printed on acid-free paper.
♾

Printed in the United States of America
Set in Monotype Sabon

Dedicated to the memory of Karl Popper, Hugh Everett and Alan Turing, and to Richard Dawkins. This book takes their ideas seriously.

Contents

Preface

If there is a single motivation for the world-view set out in this book, it is that thanks largely to a succession of extraordinary scientific discoveries, we now possess some extremely deep theories about the structure of reality. If we are to understand the world on more than a superficial level, it must be through those theories and through reason, and not through our preconceptions, received opinion or even common sense. Our best theories are not only truer than common sense, they make far more sense than common sense does. We must take them seriously, not merely as pragmatic foundations for their respective fields but as explanations of the world. And I believe that we can achieve the greatest understanding if we consider them not singly but jointly, for they are inextricably related.

It may seem odd that this suggestion – that we should try to form a rational and coherent world-view on the basis of our best, most fundamental theories – should be at all novel or controversial. Yet in practice it is. One reason is that each of these theories has, when it is taken seriously, very counter-intuitive implications. Consequently, all sorts of attempts have been made to avoid facing those implications, by making *ad hoc* modifications or reinterpretations of the theories, or by arbitrarily narrowing their domain of applicability, or simply by using them in practice but drawing no wider conclusions from them. I shall criticize some of these attempts (none of which, I believe, has much merit), but only when this happens to be a convenient way of explaining the theories themselves. For this book is not primarily a defence of these theories: it is an investigation of what the fabric of reality would be like if they were true.

Acknowledgements

The development of the ideas in this book was greatly assisted by conversations with Bryce DeWitt, Artur Ekert, Michael Lockwood, Enrico Rodrigo, Dennis Sciama, Frank Tipler, John Wheeler and Kolya Wolf.

I am grateful to my friends and colleagues Ruth Chang, Artur Ekert, David Johnson-Davies, Michael Lockwood, Enrico Rodrigo and Kolya Wolf, to my mother Tikvah Deutsch, and to my editors Caroline Knight and Ravi Mirchandani (of Penguin Books) and John Woodruff, and especially to Sarah Lawrence, for their thorough, critical reading of earlier drafts of this book, and for suggesting many corrections and improvements. I am also grateful to those who have read and commented on parts of the manuscript, including Harvey Brown, Steve Graham, Rossella Lupaccini, Svein Olav Nyberg, Oliver and Harriet Strimpel, and especially Richard Dawkins and Frank Tipler.

1

The Theory of Everything

I remember being told, when I was a small child, that in ancient times it was still possible for a very learned person to know *everything that was known*. I was also told that nowadays so much is known that no one could conceivably learn more than a tiny fraction of it, even in a long lifetime. The latter proposition surprised and disappointed me. In fact, I refused to believe it. I did not know how to justify my disbelief. But I knew that I did not want things to be like that, and I envied the ancient scholars.

It was not that I wanted to memorize all the facts that were listed in the world's encyclopaedias: on the contrary, I hated memorizing facts. That is not the sense in which I expected it to be possible to know everything that was known. It would not have disappointed me to be told that more publications appear every day than anyone could read in a lifetime, or that there are 600,000 known species of beetle. I had no wish to track the fall of every sparrow. Nor did I imagine that an ancient scholar who supposedly knew everything that was known would have known everything of that sort. I had in mind a more discriminating idea of what should count as being known. By 'known', I meant *understood*.

The idea that one person might understand everything that is understood may still seem fantastic, but it is distinctly less fantastic than the idea that one person could memorize every known fact. For example, no one could possibly memorize all known observational data on even so narrow a subject as the motions of the planets, but many astronomers *understand* those motions to the full extent that they are understood. This is possible because understanding does not depend on knowing a lot of facts as such, but on having the right concepts, explanations and theories. One comparatively simple and comprehensible theory can cover an infinity of indigestible facts. Our best theory of planetary motions is Einstein's *general theory of relativity*, which early in the twentieth century superseded Newton's theories of gravity and motion. It correctly predicts, in principle, not only all planetary motions but also all other effects of gravity to the limits of accuracy of our best measurements. For a theory to predict something 'in principle' means that the predictions follow logically from the theory, even if in practice the amount of computation that would be needed to generate some of the predictions is too large to be technologically feasible, or even too large for it to be physically possible for us to carry it out in the universe as we find it.

Being able to predict things or to describe them, however accurately, is not at all the same thing as understanding them. Predictions and descriptions in physics are often expressed as mathematical formulae. Suppose that I memorize the formula from which I could, if I had the time and the inclination, calculate any planetary position that has been recorded in the astronomical archives. What exactly have I gained, compared with memorizing those archives directly? The formula is easier to remember – but then, looking a number up in the archives may be even easier than calculating it from the formula. The real advantage of the formula is that it can be used in an infinity of cases beyond the archived data, for instance to predict the results of future observations. It may also yield the historical positions of the planets more accurately, because the archived data contain observational errors. Yet even though the formula summarizes infinitely more facts than the archives do, knowing it does not amount to understanding planetary motions. Facts cannot be understood just by being summarized in a formula, any more than by being listed on paper or committed to memory. They can be understood only by being explained. Fortunately, our best theories embody deep explanations as well as accurate predictions. For example, the general theory of relativity explains gravity in terms of a new, four-dimensional geometry of curved space and time. It explains precisely how this geometry affects and is affected by matter. That explanation is the entire content of the theory; predictions about planetary motions are merely some of the consequences that we can deduce from the explanation.

What makes the general theory of relativity so important is not that it can predict planetary motions a shade more accurately than Newton's theory can, but that it reveals and explains previously unsuspected aspects of reality, such as the curvature of space and time. This is typical of scientific explanation. Scientific theories explain the objects and phenomena of our experience in terms of an underlying reality which we do not experience directly. But the ability of a theory to explain what we experience is not its most valuable attribute. Its most valuable attribute is that it explains the fabric of reality itself. As we shall see, one of the most valuable, significant and also useful attributes of human thought generally is its ability to reveal and explain the fabric of reality.

Yet some philosophers – and even some scientists – disparage the role of explanation in science. To them, the basic purpose of a scientific theory is not to explain anything, but to predict the outcomes of experiments: its entire content lies in its predictive formulae. They consider that any consistent explanation that a theory may give for its predictions is as good as any other – or as good as no explanation at all – so long as the predictions are true. This view is called *instrumentalism* (because it says that a theory is no more than an 'instrument' for making predictions). To instrumentalists, the idea that science can enable us to understand the underlying reality that accounts for our observations is a fallacy and a conceit. They do not see how anything a scientific theory may say beyond predicting the outcomes of experiments can be more than empty words. Explanations, in particular, they regard as mere psychological props: a sort of fiction which we incorporate in theories to make them more easily remembered and entertaining. The Nobel prize-winning physicist Steven Weinberg was in instrumentalist mood when he made the following extraordinary comment about Einstein's explanation of gravity:

The important thing is to be able to make predictions about images on the astronomers' photographic plates, frequencies of spectral lines, and so on, and it simply doesn't matter whether we ascribe these predictions to the physical effects of gravitational fields on the motion of planets and photons [as in pre-Einsteinian physics] or to a curvature of space and time. (*Gravitation and Cosmology*, p. 147)

Weinberg and the other instrumentalists are mistaken. What we ascribe the images on astronomers' photographic plates to *does* matter, and it matters not only to theoretical physicists like myself, whose very motivation for formulating and studying theories is the desire to understand the world better. (I am sure that this is Weinberg's motivation too: he is not really driven by an urge to predict images and spectra!) For even in purely practical applications, the explanatory power of a theory is paramount and its predictive power only supplementary. If this seems surprising, imagine that an extraterrestrial scientist has visited the Earth and given us an ultra-high-technology 'oracle' which can predict the outcome of any possible experiment, but provides no explanations. According to instrumentalists, once we had that oracle we should have no further use for scientific theories, except as a means of entertaining ourselves. But is that true? How would the oracle be used in practice? In some sense it would contain the knowledge necessary to build, say, an interstellar spaceship. But how exactly would that help us to build one, or to build another oracle of the same kind – or even a better mousetrap? The oracle only predicts the outcomes of experiments. Therefore, in order to use it at all we must first know what experiments to ask it about. If we gave it the design of a spaceship, and the details of a proposed test flight, it could tell us how the spaceship would perform on such a flight. But it could not design the spaceship for us in the first place. And even if it predicted that the spaceship we had designed would explode on take-off, it could not tell us how to prevent such an explosion. That would still be for us to work out. And before we could work it out, before we could even begin to improve the design in any way, we should have to *understand*, among other things, how the spaceship was supposed to work. Only then would we have any chance of discovering what might cause an explosion on take-off. Prediction – even perfect, universal prediction – is simply no substitute for explanation.

Similarly, in scientific research the oracle would not provide us with any new theory. Not until we already had a theory, and had thought of an experiment that would test it, could we possibly ask the oracle what would happen if the theory were subjected to that test. Thus, the oracle would not be replacing theories at all: it would be replacing experiments. It would spare us the expense of running laboratories and particle accelerators. Instead of building prototype spaceships, and risking the lives of test pilots, we could do all the testing on the ground with pilots sitting in flight simulators whose behaviour was controlled by the predictions of the oracle.

The oracle would be very useful in many situations, but its usefulness would always depend on people's ability to solve scientific problems in just the way they have to now, namely by devising explanatory theories. It would not even replace all experimentation, because its ability to predict the outcome of a particular experiment would in practice depend on how easy it was to describe the experiment accurately enough for the oracle to give a useful answer, compared with doing the experiment in reality. After all, the oracle would have to have some sort of 'user interface'. Perhaps a description of the experiment would have to be entered into it, in some standard language. In that language, some experiments would be harder to specify than others. In practice, for many experiments the specification would be too complex to be entered. Thus the oracle would have the same general advantages and disadvantages as any other source of experimental data, and it would be useful only in cases where consulting it happened to be more convenient than using other sources. To put that another way: there already is one such oracle out there, namely the physical world. It tells us the result of any possible experiment if we ask it in the right language (i.e. if we do the experiment), though in some cases it is impractical for us to 'enter a description of the experiment' in the required form (i.e. to build and operate the apparatus). But it provides no explanations.

In a few applications, for instance weather forecasting, we may be almost as satisfied with a purely predictive oracle as with an explanatory theory. But even then, that would be strictly so only if the oracle's weather forecast were complete and perfect. In practice, weather forecasts are incomplete and imperfect, and to make up for that they include explanations of how the forecasters arrived at their predictions. The explanations allow us to judge the reliability of a forecast and to deduce further predictions relevant to our own location and needs. For instance, it makes a difference to me whether today's forecast that it will be windy tomorrow is based on an expectation of a nearby high-pressure area, or of a more distant hurricane. I would take more precautions in the latter case. Meteorologists themselves also need explanatory theories about weather so that they can guess what approximations it is safe to incorporate in their computer simulations of the weather, what additional observations would allow the forecast to be more accurate and more timely, and so on.

Thus the instrumentalist ideal epitomized by our imaginary oracle, namely a scientific theory stripped of its explanatory content, would be of strictly limited utility. Let us be thankful that real scientific theories do not resemble that ideal, and that scientists in reality do not work towards that ideal.

An extreme form of instrumentalism, called *positivism* (or logical positivism), holds that all statements other than those describing or predicting observations are not only superfluous but meaningless. Although this doctrine is itself meaningless, according to its own criterion, it was nevertheless the prevailing theory of scientific knowledge during the first half of the twentieth century! Even today, instrumentalist and positivist ideas still have currency. One reason why they are superficially plausible is that, although prediction is not the purpose of science, it is part of the characteristic *method* of science. The scientific method involves postulating a new theory to explain some class of phenomena and then performing a *crucial experimental test*, an experiment for which the old theory predicts one observable outcome and the new theory another. One then rejects the theory whose predictions turn out to be false. Thus the outcome of a crucial experimental test to decide between two theories does depend on the theories' predictions, and not directly on their explanations. This is the source of the misconception that there is nothing more to a scientific theory than its predictions. But experimental testing is by no means the only process involved in the growth of scientific knowledge. The overwhelming majority of theories are rejected because they contain bad explanations, not because they fail experimental tests. We reject them without ever bothering to test them. For example, consider the theory that eating a kilogram of grass is a cure for the common cold. That theory makes experimentally testable predictions: if people tried the grass cure and found it ineffective, the theory would be proved false. But it has never been tested and probably never will be, because it contains no explanation – either of how the cure would work, or of anything else. We rightly presume it to be false. There are always infinitely many possible theories of that sort, compatible with existing observations and making new predictions, so we could never have the time or resources to test them all. What we test are new theories that seem to show promise of explaining things better than the prevailing ones do.

To say that prediction is the purpose of a scientific theory is to confuse means with ends. It is like saying that the purpose of a spaceship is to burn fuel. In fact, burning fuel is only one of many things a spaceship has to do to accomplish its real purpose, which is to transport its payload from one point in space to another. Passing experimental tests is only one of many things a theory has to do to achieve the real purpose of science, which is to explain the world.

As I have said, explanations are inevitably framed partly in terms of things we do not observe directly: atoms and forces; the interiors of stars and the rotation of galaxies; the past and the future; the laws of nature. The deeper an explanation is, the more remote from immediate experience are the entities to which it must refer.

But these entities are not fictional: on the contrary, they are part of the very fabric of reality.

Explanations often yield predictions, at least in principle. Indeed, if something is, in principle, predictable, then a sufficiently complete explanation must, in principle, make complete predictions (among other things) about it. But many intrinsically unpredictable things can also be explained and understood. For example, you cannot predict what numbers will come up on a fair (i.e. unbiased) roulette wheel. But if you understand what it is in the wheel's design and operation that makes it fair, then you can explain why predicting the numbers is impossible. And again, merely knowing that the wheel is fair is not the same as understanding what makes it fair.

It is understanding, and not mere knowing (or describing or predicting), that I am discussing. Because understanding comes through explanatory theories, and because of the generality that such theories may have, the proliferation of recorded facts does not necessarily make it more difficult to understand everything that is understood. Nevertheless most people would say – and this is in effect what was being said to me on the occasion I recalled from my childhood – that it is not only recorded facts which have been increasing at an overwhelming rate, but also the number and complexity of the theories through which we understand the world. Consequently (they say), whether or not it was ever possible for one person to understand everything that was understood at the time, it is certainly not possible now, and it is becoming less and less possible as our knowledge grows. It might seem that every time a new explanation or technique is discovered that is relevant to a given subject, another theory must be added to the list that anyone wishing to understand that subject must learn; and that when the number of such theories in any one subject becomes too great, specializations develop. Physics, for example, has split into the sciences of astrophysics, thermodynamics, particle physics, quantum field theory, and many others. Each of these is based on a theoretical framework at least as rich as the whole of physics was a hundred years ago, and many are already fragmenting into sub-specializations. The more we discover, it seems, the further and more irrevocably we are propelled into the age of the specialist, and the more remote is that hypothetical ancient time when a single person's understanding might have encompassed all that was understood.

Confronted with this vast and rapidly growing menu of the collected theories of the human race, one may be forgiven for doubting that an individual could so much as taste every dish in a lifetime, let alone, as might once have been possible, appreciate all known recipes. Yet explanation is a strange sort of food – a larger portion is not necessarily harder to swallow. A theory may be superseded by a new theory which explains more, and is more accurate, but is also easier to understand, in which case the old theory becomes redundant, and we gain more understanding while needing to learn less than before. That is what happened when Nicolaus Copernicus's theory of the Earth travelling round the Sun superseded the complex Ptolemaic system which had placed the Earth at the centre of the universe. Or a new theory may be a simplification of an existing one, as when the Arabic (decimal) notation for numbers superseded Roman numerals. (The theory here is an implicit one. Each notation renders certain operations, statements and thoughts about numbers simpler than others, and hence it embodies a theory about which relationships between numbers are useful or interesting.) Or a new theory may be a unification of two old ones, giving us more understanding than using the old ones side by side, as happened when Michael Faraday and James Clerk Maxwell unified the theories of electricity and magnetism into a single theory of electromagnetism. More indirectly, better explanations in any subject tend to improve the techniques, concepts and language with which we are trying to understand other subjects, and so our knowledge as a whole, while increasing, can become structurally more amenable to being understood.

Admittedly, it often happens that even when old theories are thus subsumed into new ones, the old ones are not entirely forgotten. Even Roman numerals are still used today for some purposes. The cumbersome methods by which people once calculated that

XIX times XVII equals CCCXXIII are never applied in earnest any more, but they are no doubt still known and understood somewhere – by historians of mathematics for instance. Does this mean that one cannot understand 'everything that is understood' without knowing Roman numerals and their arcane arithmetic? It does not. A modern mathematician who for some reason had never heard of Roman numerals would nevertheless already possess in full the understanding of their associated mathematics. By learning about Roman numerals, that mathematician would be acquiring no new understanding, only new facts – historical facts, and facts about the properties of certain arbitrarily defined symbols, rather than new knowledge about numbers themselves. It would be like a zoologist learning to translate the names of species into a foreign language, or an astrophysicist learning how different cultures group stars into constellations.

It is a separate issue whether knowing the arithmetic of Roman numerals might be necessary in the understanding of *history*. Suppose that some historical theory – some explanation – depended on the specific techniques used by the ancient Romans for multiplication (rather as, for instance, it has been conjectured that their specific plumbing techniques, based on lead pipes, which poisoned their drinking water, contributed to the decline of the Roman Empire). Then we should have to know what those techniques were if we wanted to understand history, and therefore also if we wanted to understand everything that is understood. But in the event, no current explanation of history draws upon multiplication techniques, so our records of those techniques are mere statements of facts. Everything that is understood can be understood without learning those facts. We can always look them up when, for instance, we are deciphering an ancient text that mentions them.

In continually drawing a distinction between understanding and 'mere' knowing, I do not want to understate the importance of recorded, non-explanatory information. This is of course essential to everything from the reproduction of a micro-organism (which has such information in its DNA molecules) to the most abstract human thinking. So what distinguishes understanding from mere knowing? What is an explanation, as opposed to a mere statement of fact such as a correct description or prediction? In practice, we usually recognize the difference easily enough. We know when we do not understand something, even if we can accurately describe and predict it (for instance, the course of a known disease of unknown origin), and we know when an explanation helps us to understand it better. But it is hard to give a precise definition of 'explanation' or 'understanding'. Roughly speaking, they are about 'why' rather than 'what'; about the inner workings of things; about how things really are, not just how they appear to be; about what must be so, rather than what merely happens to be so; about laws of nature rather than rules of thumb. They are also about coherence, elegance and simplicity, as opposed to arbitrariness and complexity, though none of those things is easy to define either. But in any case, understanding is one of the higher functions of the human mind and brain, and a unique one. Many other physical systems, such as animals' brains, computers and other machines, can assimilate facts and act upon them. But at present we know of nothing that is capable of understanding an explanation – or of wanting one in the first place – other than a human mind. Every discovery of a new explanation, and every act of grasping an existing explanation, depends on the uniquely human faculty of creative thought.

One can think of what happened to Roman numerals as a process of 'demotion' of an explanatory theory to a mere description of facts. Such demotions happen all the time as our knowledge grows. Originally, the Roman system of numerals did form part of the conceptual and theoretical framework through which the people who used them understood the world. But now the understanding that used to be obtained in that way is but a tiny facet of the far deeper understanding embodied in modern mathematical theories, and implicitly in modern notations.

This illustrates another attribute of understanding. It is possible to understand something without knowing that one understands it, or even without having specifically heard of it. This may sound paradoxical, but of course the whole point of deep, general explanations is that they cover unfamiliar situations as well as familiar ones. If you were a modern mathematician encountering Roman numerals for the first time, you might not instantly realize that you already understood them. You would first have to learn the facts about what they are, and then think about those facts in the light of your existing understanding of mathematics. But once you had done that, you would be able to say, in retrospect, 'Yes, there is nothing new to me in the Roman system of numerals, beyond mere facts.' And that is what it means to say that Roman numerals, in their explanatory role, are fully obsolete.

Similarly, when I say that I understand how the curvature of space and time affects the motions of planets, even in other solar systems I may never have heard of, I am not claiming that I can call to mind, without further thought, the explanation of every detail of the loops and wobbles of any planetary orbit. What I mean is that I understand the theory that contains all those explanations, and that I could therefore produce any of them in due course, given some facts about a particular planet. Having done so, I should be able to say in retrospect, 'Yes, I see nothing in the motion of that planet, other than mere facts, which is not explained by the general theory of relativity.' We understand the fabric of reality only by understanding theories that explain it. And since they explain more than we are immediately aware of, we can understand more than we are immediately aware that we understand.

I am not saying that when we understand a theory it *necessarily* follows that we understand everything it can explain. With a very deep theory, the recognition that it explains a given phenomenon may itself be a significant discovery requiring independent explanation. For example, quasars – extremely bright sources of radiation at the centre of some galaxies – were for many years one of the mysteries of astrophysics. It was once thought that new physics would be needed to explain them, but now we believe that they are explained by the general theory of relativity and other theories that were already known before quasars were discovered. We believe that quasars consist of hot matter in the process of falling into black holes (collapsed stars whose gravitational field is so intense that nothing can escape from them). Yet reaching that conclusion has required years of research, both observational and theoretical. Now that we believe we have gained a measure of understanding of quasars, we do not think that this understanding is something we already had before. Explaining quasars, albeit through existing theories, has given us genuinely new understanding. Just as it is hard to define what an explanation is, it is hard to define when a subsidiary explanation should count as an independent component of what is understood, and when it should be considered as being subsumed in the deeper theory. It is hard to define, but not so hard to recognize: as with explanations in general, in practice we know a new explanation when we are given one. Again, the difference has something to do with creativity. Explaining the motion of a particular planet, when one already understands the general explanation of gravity, is a mechanical task, though it may be a very complex one. But using existing theory to account for quasars requires creative thought. Thus, to understand everything that is understood in astrophysics today, you would have to know the theory of quasars explicitly. But you would not have to know the orbit of any specific planet.

So, even though our stock of known theories is indeed snowballing, just as our stock of recorded facts is, that still does not necessarily make the whole structure harder to understand than it used to be. For while our specific theories are becoming more numerous and more detailed, they are continually being 'demoted' as the understanding they contain is taken over by deep, general theories. And those theories are becoming fewer, deeper and more general. By 'more general' I mean that each of them says more, about a wider range of situations, than several distinct theories did previously. By 'deeper' I mean that each of them explains more – embodies more understanding – than its predecessors did, combined.

Centuries ago, if you had wanted to build a large structure such as a bridge or a cathedral you would have engaged a master builder. He would have had some knowledge of what it takes to give a structure strength and stability with the least possible expense and effort. He would not have been able to express much of this knowledge in the language of mathematics and physics, as we can today. Instead, he relied mainly on a complex collection of intuitions, habits and rules of thumb, which he had learned from his apprentice-master and then perhaps amended through guesswork and long experience. Even so, these intuitions, habits and rules of thumb were in effect *theories*, explicit and inexplicit, and they contained real knowledge of the subjects we nowadays call engineering and architecture. It was for the knowledge in those theories that you would have hired him, pitifully inaccurate though it was compared with what we have today, and of very narrow applicability. When admiring centuries-old structures, people often forget that we see only the surviving ones. The overwhelming majority of structures built in medieval and earlier times have collapsed long ago, often soon after they were built. That was especially so for innovative structures. It was taken for granted that innovation risked catastrophe, and builders seldom deviated much from designs and techniques that had been validated by long tradition. Nowadays, in contrast, it is quite rare for any structure – even one that is unlike anything that has ever been built before – to fail because of faulty design. Anything that an ancient master builder could have built, his modern colleagues can build better and with far less human effort. They can also build structures which he could hardly have dreamt of, such as skyscrapers and space stations. They can use materials which he had never heard of, such as fibreglass or reinforced concrete, and which he could hardly have used even if he could somehow have been given them, for he had only a scanty and inaccurate understanding of how materials work.

Progress to our current state of knowledge was not achieved by accumulating more theories of the same kind as the master builder knew. Our knowledge, both explicit and inexplicit, is not only much greater than his but structurally different too. As I have said, the modern theories are fewer, more general and deeper. For each situation that the master builder faced while building something in his repertoire – say, when deciding how thick to make a load-bearing wall – he had a fairly specific intuition or rule of thumb, which, however, could give hopelessly wrong answers if applied to novel situations. Today one deduces such things from a theory that is general enough for it to be applied to walls made of any material, in all situations: on the Moon, underwater, or wherever. The reason why it is so general is that it is based on quite deep explanations of how materials and structures work. To find the proper thickness of a wall that is to be made from an unfamiliar material, one uses the same theory as for any other wall, but starts the calculation by assuming different facts – by using different numerical values for the various parameters. One has to look up those facts, such as the tensile strength and elasticity of the material, but one needs no additional understanding.

That is why, despite understanding incomparably more than an ancient master builder did, a modern architect does not require a longer or more arduous training. A typical theory in a modern student's syllabus may be harder to understand than any of the master builder's rules of thumb; but the modern theories are far fewer, and their explanatory power gives them other properties such as beauty, inner logic and connections with other subjects which make them easier to learn. Some of the ancient rules of thumb are now known to be erroneous, while others are known to be true, or to be good approximations to the truth, and we know why that is so. A few are still in use. But none of them is any longer the source of anyone's understanding of what makes structures stand up.

I am not, of course, denying that specialization is occurring in many subjects in which knowledge is growing, including architecture. This is not a one-way process, for specializations often disappear too: wheels are no longer designed or made by wheelwrights, nor ploughs by ploughwrights, nor are letters written by scribes. It is nevertheless quite evident that the deepening, unifying tendency I have been describing is not the only one at work: a continual *broadening* is going on at the same time. That is, new ideas often do more than just supersede, simplify or unify existing ones. They also extend human understanding into areas that were previously not understood at all – or whose very existence was not guessed at. They may open up new opportunities, new problems, new specializations and even new subjects. And when that happens it may give us, at least temporarily, more to learn in order to understand it all.

The science of medicine is perhaps the most frequently cited case of increasing specialization seeming to follow inevitably from increasing knowledge, as new cures and better treatments for more diseases are discovered. But even in medicine the opposite, unifying tendency is also present, and is becoming stronger. Admittedly, many functions of the body are still poorly understood, and so are the mechanisms of many diseases. Consequently some areas of medical knowledge still consist mainly of collections of recorded facts, together with the skills and intuitions of doctors who have experience of particular diseases and particular treatments, and who pass on these skills and intuitions from one generation to the next. Much of medicine, in other words, is still in the rule-of-thumb era, and when new rules of thumb are discovered there is indeed more incentive for specialization. But as medical and biochemical research comes up with deeper explanations of disease processes (and healthy processes) in the body, understanding is also on the increase. More general concepts are replacing more specific ones as common, underlying molecular mechanisms are found for dissimilar diseases in different parts of the body. Once a disease can be understood as fitting into a general framework, the role of the specialist diminishes. Instead, physicians coming across an unfamiliar disease or a rare complication can rely increasingly on explanatory theories. They can look up such facts as are known. But then they may be able to apply a general theory to work out the required treatment, and expect it to be effective even if it has never been used before.

Thus the issue of whether it is becoming harder or easier to understand everything that is understood depends on the overall balance between these two opposing effects of the growth of knowledge: the increasing *breadth* of our theories, and their increasing *depth*. Breadth makes it harder; depth makes it easier. One thesis of this book is that, slowly but surely, depth is winning. In other words, the proposition that I refused to believe as a child is indeed false, and practically the opposite is true. We are not heading away from a state in which one person could understand everything that is understood, but towards it.

It is not that we shall soon understand *everything*. That is a completely different issue. I do not believe that we are now, or ever shall be, close to understanding *everything there is*. What I am discussing is the possibility of understanding *everything that is understood*. That depends more on the structure of our knowledge than on its content. But of course the structure of our knowledge – whether it is expressible in theories that fit together as a comprehensible whole – does depend on what the fabric of reality, as a whole, is like. If knowledge is to continue its open-ended growth, and if we are nevertheless heading towards a state in which one person could understand everything that is understood, then the depth of our theories must continue to grow fast enough to make this possible. That can happen only if the fabric of reality is itself highly unified, so that more and more of it can become understood as our knowledge grows. If that happens, then eventually our theories will become so general, deep and integrated with one another that they will effectively become a single theory of a unified fabric of reality. This theory will still not explain every aspect of reality: that is unattainable. But it will encompass all known explanations, and will apply to the whole fabric of reality in so far as it is understood. Whereas all previous theories related to particular subjects, this will be a theory of all subjects: a *Theory of Everything*.

It will not, of course, be the last such theory, only the first. In science we take it for granted that even our best theories are bound to be imperfect and problematic in some ways, and we expect them to be superseded in due course by deeper, more accurate theories. Such progress is not brought to a halt when we discover a universal theory. For example, Newton gave us the first universal theory of gravity and a unification of, among other things, celestial and terrestrial mechanics. But his theories have been superseded by Einstein's general theory of relativity which additionally incorporates geometry (formerly regarded as a branch of mathematics) into physics, and in so doing provides far deeper explanations as well as being more accurate. The first fully universal theory – which I shall call the Theory of Everything – will, like all our theories before and after it, be neither perfectly true nor infinitely deep, and so will eventually be superseded. But it will not be superseded through unifications with theories about other subjects, for it will already be a theory of all subjects. In the past, some great advances in understanding came about through great unifications. Others came through structural changes in the way we were understanding a particular subject – as when we ceased to think of the Earth as being the centre of the universe. After the first Theory of Everything, there will be no more great unifications. All subsequent great discoveries will take the form of changes in the way we understand the world as a whole: shifts in our world-view. The attainment of a Theory of Everything will be the last great unification, and at the same time it will be the first across-the-board shift to a new world-view. I believe that such a unification and shift are now under way. The associated world-view is the theme of this book.

I must stress immediately that I am not referring merely to the 'theory of everything' which some particle physicists hope they will soon discover. *Their* 'theory of everything' would be a unified theory of all the basic forces known to physics, namely gravity, electromagnetism and nuclear forces. It would also describe all the types of subatomic particles that exist, their masses, spins, electric charges and other properties, and how they interact. Given a sufficiently precise description of the initial state of any isolated physical system, it would in principle predict the future behaviour of the system. Where the exact behaviour of a system was intrinsically unpredictable, it would describe all possible behaviours and predict their probabilities. In practice, the initial states of interesting systems often cannot be ascertained very accurately, and in any case the calculation of the predictions would be too complicated to be carried out in all but the simplest cases. Nevertheless, such a unified theory of particles and forces, together with a specification of the initial state of the universe at the Big Bang (the violent explosion with which the universe began), would in principle contain all the information necessary to predict everything that can be predicted (Figure 1.1).

But prediction is not explanation. The hoped-for 'theory of everything', even if combined with a theory of the initial state, will at best provide only a tiny facet of a real Theory of Everything. It may *predict* everything (in principle). But it cannot be expected to *explain* much more than existing theories do, except for a few phenomena that are dominated by the nuances of subatomic interactions, such as collisions inside particle accelerators, and the exotic history of particle transmutations in the Big Bang. What motivates the use of the term 'theory of everything' for such a narrow, albeit fascinating, piece of knowledge? It is, I think, another mistaken view of the nature of science, held disapprovingly by many critics of science and (alas) approvingly by many scientists, namely that science is essentially *reductionist*. That is to say, science allegedly explains things reductively – by analysing them into components. For example, the resistance of a wall to being penetrated or knocked down is explained by regarding the wall as a vast aggregation of interacting molecules. The properties of those molecules are themselves explained in terms of their constituent atoms, and the interactions of these atoms with one another, and so on down to the smallest particles and most basic forces. Reductionists think that all scientific explanations, and perhaps all sufficiently deep explanations of any kind, take that form.

The reductionist conception leads naturally to a classification of

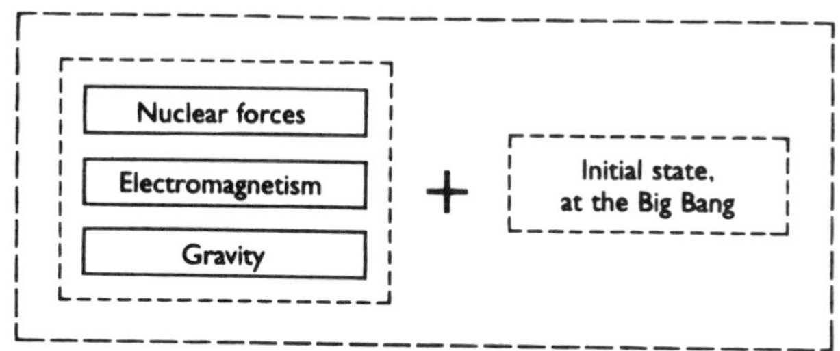

FIGURE 1.1 *An inadequate conception of the 'theory of everything'.*

ubjects and theories in a hierarchy, according to how close they are to the 'lowest-level' predictive theories that are known. In this hierarchy, logic and mathematics form the immovable bedrock on which the edifice of science is built. The foundation stone would be a reductive 'theory of everything', a universal theory of particles, forces, space and time, together with some theory of what the initial state of the universe was. The rest of physics forms the first few storeys. Astrophysics and chemistry are at a higher level, geology even higher, and so on. The edifice branches into many towers of increasingly high-level subjects like biochemistry, biology and genetics. Perched at the tottering, stratospheric tops are subjects like the theory of evolution, economics, psychology and computer science, which in this picture are almost inconceivably derivative.

At present, we have only approximations to a reductive 'theory of everything'. These can already predict quite accurate laws of motion for individual subatomic particles. From these laws, present-day computers can calculate the motion of any isolated group of a few interacting particles in some detail, given their initial state. But even the smallest speck of matter visible to the naked eye contains trillions of atoms, each composed of many subatomic particles, and is continually interacting with the outside world; so it is quite infeasible to predict its behaviour particle by particle. By supplementing the exact laws of motion with various approximation schemes, we can predict some aspects of the gross behaviour of quite large objects – for instance, the temperature at which a given chemical compound will melt or boil. Much of basic chemistry has been reduced to physics in this way. But for higher-level sciences the reductionist programme is a matter of principle only. No one expects actually to deduce many principles of biology, psychology or politics from those of physics. The reason why higher-level subjects can be studied at all is that under special circumstances the stupendously complex behaviour of vast numbers of particles resolves itself into a measure of simplicity and comprehensibility. This is called *emergence*: high-level simplicity 'emerges' from low-level complexity. High-level phenomena about which there are comprehensible facts that are not simply deducible from lower-level theories are called *emergent phenomena*. For example, a wall might be strong because its builders feared that their enemies might try to force their way through it. This is a high-level explanation of the wall's strength, not deducible from (though not incompatible with) the low-level explanation I gave above. 'Builders', 'enemies', 'fear' and 'trying' are all emergent phenomena. The purpose of high-level sciences is to enable us to understand emergent phenomena, of which the most important are, as we shall see, *life*, *thought* and *computation*.

By the way, the opposite of reductionism, *holism* – the idea that the only legitimate explanations are in terms of higher-level systems – is an even greater error than reductionism. What do holists expect us to do? Cease our search for the molecular origin of diseases? Deny that human beings are made of subatomic particles? Where reductive explanations exist, they are just as desirable as any other explanations. Where whole sciences are reducible to lower-level sciences, it is just as incumbent upon us as scientists to find those reductions as it is to discover any other knowledge.

A reductionist thinks that science is about analysing things into components. An instrumentalist thinks that it is about predicting things. To either of them, the existence of high-level sciences is merely a matter of convenience. Complexity prevents us from using fundamental physics to make high-level predictions, so instead we guess what those predictions would be if we could make them – emergence gives us a chance of doing that successfully – and supposedly that is what the higher-level sciences are about. Thus to reductionists and instrumentalists, who disregard both the real structure and the real purpose of scientific knowledge, the base of the predictive hierarchy of physics is by definition the 'theory of everything'. But to everyone else scientific knowledge consists of explanations, and the structure of scientific explanation does not reflect the reductionist hierarchy. There are explanations at every level of the hierarchy. Many of them are autonomous, referring only to concepts at that particular level (for instance, 'the bear ate the honey because it was hungry'). Many involve deductions in the opposite direction to that of reductive explanation. That is, they explain things not by analysing them into smaller, simpler things but by regarding them as components of larger, more complex things – about which we nevertheless have explanatory theories. For example, consider one particular copper atom at the tip of the nose of the statue of Sir Winston Churchill that stands in Parliament Square in London. Let me try to explain why that copper atom is there. It is because Churchill served as prime minister in the House of Commons nearby; and because his ideas and leadership contributed to the Allied victory in the Second World War; and because it is customary to honour such people by putting up statues of them; and because bronze, a traditional material for such statues, contains copper, and so on. Thus we explain a low-level physical observation – the presence of a copper atom at a particular location – through extremely high-level theories about emergent phenomena such as ideas, leadership, war and tradition.

There is no reason why there should exist, even in principle, any lower-level *explanation* of the presence of that copper atom than the one I have just given. Presumably a reductive 'theory of everything' would in principle make a low-level *prediction* of the probability that such a statue will exist, given the condition of (say) the solar system at some earlier date. It would also in principle describe how the statue probably got there. But such descriptions and predictions (wildly infeasible, of course) would explain nothing. They would merely describe the trajectory that each copper atom followed from the copper mine, through the smelter and the sculptor's studio, and so on. They could also state how those trajectories were influenced by forces exerted by surrounding atoms, such as those comprising the miners' and sculptor's bodies, and so predict the existence and shape of the statue. In fact such a prediction would have to refer to atoms all over the planet, engaged in the complex motion we call the Second World War, among other things. But even if you had the superhuman capacity to follow such lengthy predictions of the copper atom's being there, you would still not be able to say, 'Ah yes, now I understand why it is there.' You would merely know that its arrival there in that way was inevitable (or likely, or whatever), given all the atoms' initial configurations and the laws of physics. If you wanted to understand why, you would still have no option but to take a further step. You would have to inquire into what it was about that configuration of atoms, and those trajectories, that gave them the propensity to deposit a copper atom at this location. Pursuing this inquiry would be a creative task, as discovering new explanations always is. You would have to discover that certain atomic configurations support emergent phenomena such as leadership and war, which are related to one another by high-level explanatory theories. Only when you knew those theories could you understand fully why that copper atom is where it is.

In the reductionist world-view, the laws governing subatomic particle interactions are of paramount importance, as they are the base of the hierarchy of all knowledge. But in the real structure of scientific knowledge, and in the structure of our knowledge generally, such laws have a much more humble role.

What is that role? It seems to me that none of the candidates for a 'theory of everything' that has yet been contemplated contains much that is new by way of explanation. Perhaps the most innovative approach from the explanatory point of view is *superstring theory*, in which extended objects, 'strings', rather than point-like particles, are the elementary building blocks of matter. But no existing approach offers an entirely new mode of explanation – new in the sense of Einstein's explanation of gravitational forces in terms of curved space and time. In fact, the 'theory of everything' is expected to inherit virtually its entire explanatory structure – its physical concepts, its language, its mathematical formalism and the form of its explanations – from the existing theories of electromagnetism, nuclear forces and gravity. Therefore we may look to this underlying structure, which we already know from existing theories, for the contribution of fundamental physics to our overall understanding.

There are two theories in physics which are considerably deeper than all others. The first is the general theory of relativity, which as I have said is our best theory of space, time and gravity. The second, *quantum theory*, is even deeper. Between them, these two theories (and not any existing or currently envisaged theory of subatomic particles) provide the detailed explanatory and formal framework within which all other theories in modern physics are expressed, and they contain overarching physical principles to which all other theories conform. A unification of general relativity and quantum theory – to give a *quantum theory of gravity* – has been a major quest of theoretical physicists for several decades, and would have to form part of any theory of everything in either the narrow or the broad sense of the term. As we shall see in the next chapter, quantum theory, like relativity, provides a revolutionary new mode of explanation of physical reality. The reason why quantum theory is the deeper of the two lies more outside physics than within it, for its ramifications are very wide, extending far beyond physics – and even beyond science itself as it is normally conceived. Quantum theory is one of what I shall call the *four main strands* of which our current understanding of the fabric of reality is composed.

Before I say what the other three strands are, I must mention another way in which reductionism misrepresents the structure of scientific knowledge. Not only does it assume that explanation always consists of analysing a system into smaller, simpler systems, it also assumes that all explanation is of later events in terms of earlier events; in other words, that the only way of explaining something is to state its *causes*. And this implies that the earlier the events in terms of which we explain something, the better the explanation, so that ultimately the best explanations of all are in terms of the initial state of the universe.

A 'theory of everything' which excludes a specification of the initial state of the universe is not a complete description of physical reality because it provides only laws of motion; and laws of motion, by themselves, make only conditional predictions. That is, they never state categorically what happens, but only what will happen at one time given what was happening at another time. Only if a complete specification of the initial state is provided can a complete description of physical reality in principle be deduced. Current cosmological theories do not provide a complete specification of the initial state, even in principle, but they do say that the universe was initially very small, very hot and very uniform in structure. We also know that it cannot have been perfectly uniform because that would be incompatible, according to the theory, with the distribution of galaxies we observe across the sky today. The initial variations in density, 'lumpiness', would have been greatly enhanced by gravitational clumping (that is, relatively dense regions would have attracted more matter and become denser), so they need only have been very slight initially. But, slight though they were, they are of the greatest significance in any reductionist description of reality, because almost everything that we see happening around us, from the distribution of stars and galaxies in the sky to the appearance of bronze statues on planet Earth, is, from the point of view of fundamental physics, a consequence of those variations. If our reductionist description is to cover anything more than the grossest features of the observed universe, we need a theory specifying those all-important initial deviations from uniformity.

Let me try to restate this requirement without the reductionist bias. The laws of motion for any physical system make only conditional predictions, and are therefore compatible with many possible histories of that system. (This issue is independent of the limitations on predictability that are imposed by quantum theory, which I shall discuss in the next chapter.) For instance, the laws of motion governing a cannon-ball fired from a gun are compatible with many possible trajectories, one for every possible direction and elevation in which the gun could have been pointing when it was fired (Figure 1.2). Mathematically, the laws of motion can be expressed as a set of equations called the *equations of motion*. These have many different solutions, one describing each possible trajectory. To specify which solution describes the actual trajectory, we must provide *supplementary data* – some data about what actually happens. One way of doing that is to specify the initial state, in this case the direction in which the gun was pointing. But there are other ways too. For example, we could just as well specify the final state – the position and direction of motion of the cannon-

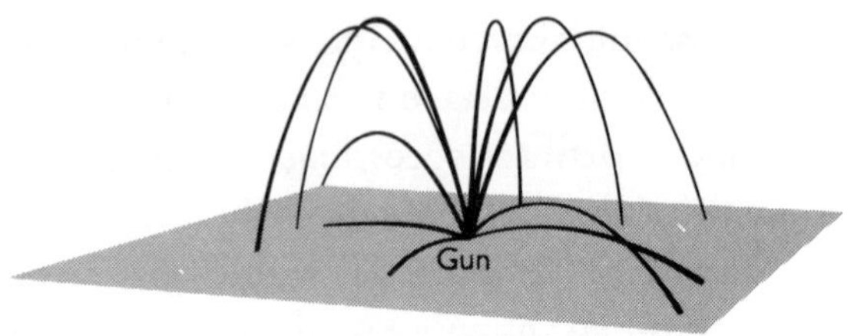

FIGURE 1.2 *Some possible trajectories of a cannon-ball fired from a gun. Each trajectory is compatible with the laws of motion, but only one of them is the trajectory on a particular occasion.*

ball at the moment it lands. Or we could specify the position of the highest point of the trajectory. It does not matter what supplementary data we give, so long as they pick out one particular solution of the equations of motion. The combination of any such supplementary data with the laws of motion amounts to a theory that describes everything that happens to the cannon-ball between firing and impact.

Similarly, the laws of motion for physical reality as a whole would have many solutions, each corresponding to a distinct history. To complete the description, we should have to specify which history is the one that has actually occurred, by giving enough supplementary data to yield one of the many solutions of the equations of motion. In simple cosmological models at least, one way of giving such data is to specify the initial state of the universe. But alternatively we could specify the final state, or the state at any other time; or we could give some information about the initial state, some about the final state, and some about states in between. In general, the combination of enough supplementary data of any sort with the laws of motion would amount to a complete description, in principle, of physical reality.

For the cannon-ball, once we have specified, say, the final state it is straightforward to calculate the initial state, and vice versa, so there is no practical difference between different methods of specifying the supplementary data. But for the universe most such calculations are intractable. I have said that we infer the existence of 'lumpiness' in the initial conditions from observations of 'lumpiness' today. But that is exceptional: most of our knowledge of supplementary data – of what specifically happens – is in the form of high-level theories about emergent phenomena, and is therefore by definition not practically expressible in the form of statements about the initial state. For example, in most solutions of the equations of motion the initial state of the universe does not have the right properties for life to evolve from it. Therefore our knowledge that life *has* evolved is a significant piece of the supplementary data. We may never know what, specifically, this restriction implies about the detailed structure of the Big Bang, but we can draw conclusions from it directly. For example, the earliest accurate estimate of the age of the Earth was made on the basis of the biological theory of evolution, contradicting the best physics of the day. Only a reductionist prejudice could make us feel that this was somehow a less valid form of reasoning, or that in general it is more 'fundamental' to theorize about the initial state than about emergent features of reality.

Even in the domain of fundamental physics, the idea that theories of the initial state contain our deepest knowledge is a serious misconception. One reason is that it logically excludes the possibility of explaining the initial state itself – why the initial state was what it was – but in fact we have explanations of many aspects of the initial state. And more generally, no theory of *time* can possibly explain it in terms of anything 'earlier'; yet we do have deep explanations, from general relativity and even more from quantum theory, of the nature of time (see Chapter 11).

Thus the character of many of our descriptions, predictions and explanations of reality bear no resemblance to the 'initial state plus laws of motion' picture that reductionism leads to. There is no reason to regard high-level theories as in any way 'second-class citizens'. Our theories of subatomic physics, and even of quantum theory or relativity, are in no way privileged relative to theories about emergent properties. None of these areas of knowledge can possibly subsume all the others. Each of them has logical implications for the others, but not all the implications can be stated, for they are emergent properties of the other theories' domains. In fact, the very terms 'high level' and 'low level' are misnomers. The laws of biology, say, are high-level, emergent consequences of the laws of physics. But logically, some of the laws of physics are then 'emergent' consequences of the laws of biology. It could even be that, between them, the laws governing biological and other emergent phenomena would entirely determine the laws of fundamental physics. But in any case, when two theories are logically related, logic does not dictate which of them we ought to regard as determining, wholly or partly, the other. That depends on the explanatory relationships between the theories. The truly privileged theories are not the ones referring to any particular scale of size or complexity, nor the ones situated at any particular level of the predictive hierarchy – but the ones that contain the deepest explanations. The fabric of reality does not consist only of reductionist ingredients like space, time and subatomic particles, but also of life, thought, computation and the other things to which those explanations refer. What makes a theory more fundamental, and less derivative, is not its closeness to the supposed predictive base of physics, but its closeness to our deepest explanatory theories.

Quantum theory is, as I have said, one such theory. But the other three main strands of explanation through which we seek to understand the fabric of reality are all 'high level' from the point of view of quantum physics. They are the *theory of evolution* (primarily the evolution of living organisms), *epistemology* (the theory of knowledge) and the *theory of computation* (about computers and what they can and cannot, in principle, compute). As I shall show, such deep and diverse connections have been discovered between the basic principles of these four apparently independent subjects that it has become impossible to reach our best understanding of any one of them without also understanding the other three. The four of them taken together form a coherent explanatory structure that is so far-reaching, and has come to encompass so much of our understanding of the world, that in my view it may already properly be called the first real Theory of Everything. Thus we have arrived at a significant moment in the history of ideas – the moment when the scope of our understanding begins to be fully universal. Up to now, all our understanding has been about some aspect of reality, untypical of the whole. In the future it will be about a unified conception of reality: all explanations will be understood against the backdrop of universality, and every new idea will automatically tend to illuminate not just a particular subject, but, to varying degrees, all subjects. The dividend of understanding that we shall eventually reap from this last great unification may far surpass that yielded by any previous one. For we shall see that it is not only physics that is being unified and explained here, and not only science, but also potentially the far reaches of philosophy, logic and mathematics, ethics, politics and aesthetics; perhaps everything that we currently understand, and probably much that we do not yet understand.

What conclusion, then, would I address to my younger self, who rejected the proposition that the growth of knowledge was making the world ever less comprehensible? I would agree with him, though I now think that the important issue is not really whether what our particular species understands can be understood by *one* of its members. It is whether the fabric of reality itself is truly unified and comprehensible. There is every reason to believe that it is. As a child, I merely knew this; now I can explain it.

TERMINOLOGY

epistemology The study of the nature of knowledge and the processes that create it.

explanation (roughly) A statement about the nature of things and the reasons for things.

instrumentalism The view that the purpose of a scientific theory is to predict the outcomes of experiments.

positivism An extreme form of instrumentalism which holds that all statements other than those describing or predicting observations are meaningless. (This view is itself meaningless according to its own criterion.)

reductive A reductive explanation is one that works by analysing things into lower-level components.

reductionism The view that scientific explanations are inherently reductive.

holism The idea that the only legitimate explanations are in terms of higher-level systems; the opposite of reductionism.

emergence An emergent phenomenon is one (such as life, thought or computation) about which there are comprehensible facts or explanations that are not simply deducible from lower-level theories, but which may be explicable or predictable by higher-level theories referring directly to that phenomenon.

SUMMARY

Scientific knowledge, like all human knowledge, consists primarily of explanations. Mere facts can be looked up, and predictions are important only for conducting crucial experimental tests to discriminate between competing scientific theories that have already passed the test of being good explanations. As new theories supersede old ones, our knowledge is becoming both broader (as new subjects are created) and deeper (as our fundamental theories explain more, and become more general). Depth is winning. Thus we are not heading away from a state in which one person could understand everything that was understood, but towards it. Our deepest theories are becoming so integrated with one another that they can be understood only jointly, as a single theory of a unified fabric of reality. This Theory of Everything has a far wider scope than the 'theory of everything' that elementary particle physicists are seeking, because the fabric of reality does not consist only of reductionist ingredients such as space, time and subatomic particles, but also, for example, of life, thought and computation. The *four main strands* of explanation which may constitute the first Theory of Everything are:

quantum physics Chapters 2, 9, 11, 12, 13, 14

epistemology Chapters 3, 4, 7, 10, 13, 14

the theory of computation Chapters 5, 6, 9, 10, 13, 14

the theory of evolution Chapters 8, 13, 14.

The next chapter is about the first and most important of the four strands, quantum physics.

2

Shadows

There is no better, there is no more open door by which you can enter into the study of natural philosophy, than by considering the physical phenomena of a candle.

Michael Faraday (*A Course of Six Lectures on the Chemical History of a Candle*)

In his popular Royal Institution lectures on science, Michael Faraday used to urge his audiences to learn about the world by considering what happens when a candle burns. I am going to consider an electric torch (or flashlight) instead. This is quite fitting, for much of the technology of an electric torch is based on Faraday's discoveries.

I am going to describe some experiments which demonstrate phenomena that are at the core of quantum physics. Experiments of this sort, with many variations and refinements, have been the bread and butter of quantum optics for many years. There is no controversy about the results, yet even now some of them are hard to believe. The basic experiments are remarkably austere. They require neither specialized scientific instruments nor any great knowledge of mathematics or physics – essentially, they involve nothing but casting shadows. But the patterns of light and shadow that an ordinary electric torch can cast are very strange. When considered carefully they have extraordinary ramifications. Explaining them requires not just new physical laws but a new *level* of description and explanation that goes beyond what was previously regarded as being the scope of science. But first, it reveals the existence of parallel universes. How can it? What conceivable pattern of shadows could have implications like that?

Imagine an electric torch switched on in an otherwise dark room. Light emanates from the filament of the torch's bulb and fills out part of a cone. In order not to complicate the experiment with reflected light, the walls of the room should be totally absorbent, matt black. Alternatively, since we are only imagining these experiments, we could imagine a room of astronomical size, so that there is no time for any light to reach the walls and return before the experiment is completed. Figure 2.1 illustrates the situation. But it is somewhat misleading: if we were observing the torch from the side we should be able to see neither it nor, of course, its light. Invisibility is one of the more straightforward properties of light. We see light only if it enters our eyes (though we usually speak of seeing the object in our line of sight that last affected that light).

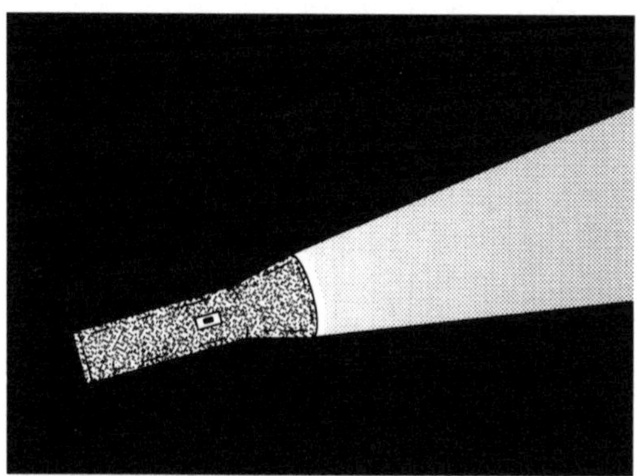

FIGURE 2.1 *Light from an electric torch (flashlight).*

We cannot see light that is just passing by. If there were a reflective object in the beam, or even some dust or water droplets to scatter the light, we could see where it was. But there is nothing in the beam, and we are observing from outside it, so none of its light reaches us. An accurate representation of what we should see would be a completely black picture. If there were a second source of light we might be able to see the torch, but still not its light. Beams of light, even the most intense light that we can generate (from lasers), pass through each other as if nothing were there at all.

Figure 2.1 does show that the light is brightest near the torch, and gets dimmer farther away as the beam spreads out to illuminate an ever larger area. To an observer within the beam, backing steadily away from the torch, the reflector would appear ever smaller and then, when it could only be seen as a single point, ever fainter. Or would it? Can light really be spread more and more thinly without limit? The answer is no. At a distance of approximately ten thousand kilometres from the torch, its light would be too faint for the human eye to detect and the observer would see nothing. That is, a human observer would see nothing; but what about an animal with more sensitive vision? Frogs' eyes are several times more sensitive than human eyes – just enough to make a significant difference in this experiment. If the observer were a frog, and it kept moving ever farther away from the torch, the moment at which it entirely lost sight of the torch would never come. Instead, the frog would see the torch begin to flicker. The flickers would come at irregular intervals that would become longer as the frog moved farther away. But the brightness of the individual flickers would not diminish. At a distance of one hundred million kilometres from the torch, the frog would see on average only one flicker of light per day, but that flicker would be as bright as any that it observed at any other distance.

Frogs cannot tell us what they see. So in real experiments we use photomultipliers (light detectors which are even more sensitive than frogs' eyes), and we thin out the light by passing it through dark filters, rather than by observing it from a hundred million kilometres away. But the principle is the same, and so is the result: neither apparent darkness nor uniform dimness, but flickering, with the individual flickers equally bright no matter how dark a filter we use. This flickering indicates that there is a limit to how thinly light can be evenly spread. Borrowing the terminology of goldsmiths, one might say that light is not infinitely 'malleable'. Like gold, a small amount of light can be evenly spread over a very large area, but eventually if one tries to spread it out further it gets lumpy. Even if gold atoms could somehow be prevented from clumping together, there is a point beyond which they cannot be subdivided without ceasing to be gold. So the only way in which one can make a one-atom-thick gold sheet even thinner is to space the atoms farther apart, with empty space between them. When they are sufficiently far apart it becomes misleading to think of them as forming a continuous sheet. For example, if each gold atom were on average several centimetres from its nearest neighbour, one might pass one's hand through the 'sheet' without touching any gold at all. Similarly, there is an ultimate lump or 'atom' of light, a *photon*. Each flicker seen by the frog is caused by a photon striking the retina of its eye. What happens when a beam of light gets fainter is not that the photons themselves get fainter, but that they get farther apart, with empty space between them (Figure 2.2). When the beam is very faint it can be misleading to call it a 'beam', for it is not continuous. During periods when the frog sees nothing it is not because the light entering its eye is too weak to affect the retina, but because no light has entered its eye at all.

This property of appearing only in lumps of discrete sizes is called *quantization*. An individual lump, such as a photon, is called a *quantum* (plural *quanta*). Quantum theory gets its name from this property, which it attributes to all measurable physical quantities – not just to things like the amount of light, or the mass of gold, which

FIGURE 2.2 *Frogs can see individual photons.*

are quantized because the entities concerned, though apparently continuous, are really made of particles. Even for quantities like distance (between two atoms, say), the notion of a continuous range of possible values turns out to be an idealization. There are no measurable continuous quantities in physics. There are many new effects in quantum physics, and on the face of it quantization is one of the tamest, as we shall see. Yet in a sense it remains the key to all the others, for if everything is quantized, how does any quantity change from one value to another? How does any object get from one *place* to another if there is not a continuous range of intermediate places for it to be on the way? I shall explain how in Chapter 9, but let me set that question aside for the moment and return to the vicinity of the torch, where the beam looks continuous because every second it pours about 10^{14} (a hundred trillion) photons into an eye that looks into it.

Is the boundary between the light and the shadow perfectly sharp, or is there a grey area? There is usually a fairly wide grey area, and one reason for this is shown in Figure 2.3. There is a dark region (called the *umbra*) where light from the filament cannot reach. There is a bright region which can receive light from anywhere on the filament. And because the filament is not a geometrical point, but has a certain size, there is also a *penumbra* between the bright and dark regions: a region which can receive light from some parts of the filament but not from others. If one observes from within the penumbra, one can see only part of the filament and the illumination is less there than in the fully illuminated, bright region.

However, the size of the filament is not the only reason why real torchlight casts penumbras. The light is affected in all sorts of other ways by the reflector behind the bulb, by the glass front of the torch, by various seams and imperfections, and so on. So we expect quite a complicated pattern of light and shadow from a real torch, just because the torch itself is quite complicated. But the incidental properties of torches are not the subject of these experiments. Behind our question about torchlight there is a more fundamental question about light in general: is there, in principle, any limit on how sharp a shadow can be (in other words, on how narrow a

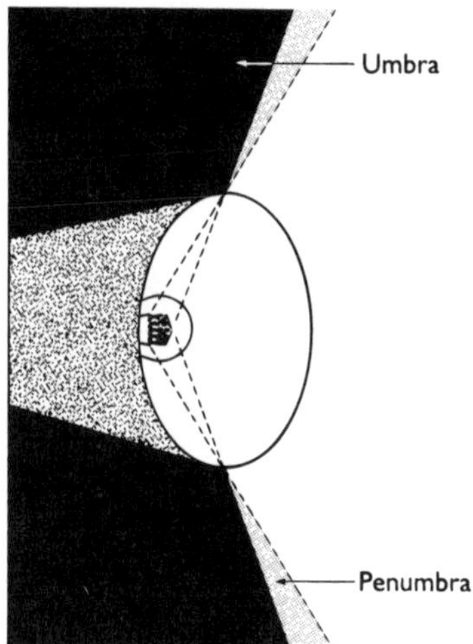

FIGURE 2.3 *The umbra and penumbra of a shadow.*

penumbra can be)? For instance, if the torch were made of perfectly black (non-reflecting) material, and if one were to use smaller and smaller filaments, could one then make the penumbra narrower and narrower, without limit?

Figure 2.3 makes it look as though one could: if the filament had no size, there would be no penumbra. But in drawing Figure 2.3 I have made an assumption about light, namely that it travels only in straight lines. From everyday experience we know that it does, for we cannot see round corners. But careful experiments show that light does not always travel in straight lines. Under some circumstances it bends.

This is hard to demonstrate with a torch alone, just because it is difficult to make very tiny filaments and very black surfaces. These practical difficulties mask the limits that fundamental physics imposes on the sharpness of shadows. Fortunately, the bending of light can also be demonstrated in a different way. Suppose that the light of a torch passes through two successive small holes in otherwise opaque screens, as shown in Figure 2.4, and that the emerging light falls on a third screen beyond. Our question now is this: if the experiment is repeated with ever smaller holes and with ever

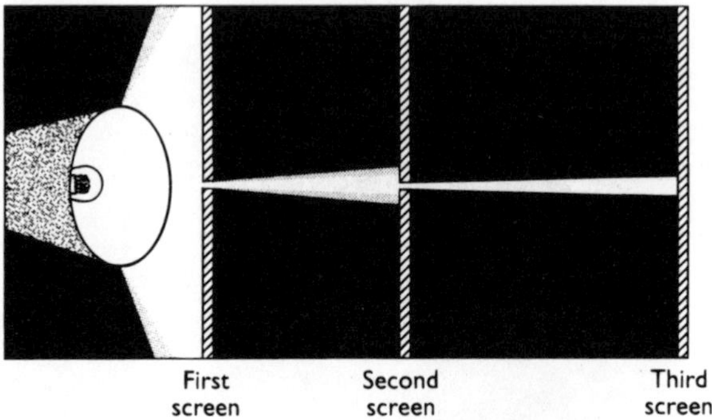

FIGURE 2.4 *Making a narrow beam by passing light through two successive holes.*

greater separation between the first and second screens, can one bring the umbra – the region of total darkness – ever closer, without limit, to the straight line through the centres of the two holes? Can the illuminated region between the second and third screens be confined to an arbitrarily narrow cone? In goldsmiths' terminology, we are now asking something like 'how "ductile" is light' – how fine a thread can it be drawn into? Gold can be drawn into threads one ten-thousandth of a millimetre thick.

It turns out that light is not as ductile as gold! Long before the holes get as small as a ten-thousandth of a millimetre, in fact even with holes as large as a millimetre or so in diameter, the light begins noticeably to rebel. Instead of passing through the holes in straight lines, it refuses to be confined and spreads out after each hole. And as it spreads, it 'frays'. The smaller the hole is, the more the light spreads out from its straight-line path. Intricate patterns of light and shadow appear. We no longer see simply a bright region and a dark region on the third screen, with a penumbra in between, but instead concentric rings of varying thickness and brightness. There is also colour, because white light consists of a mixture of photons of various colours, and each colour spreads and frays in a slightly different pattern. Figure 2.5 shows a typical pattern that might be formed on the third screen by white light that has passed through holes in the first two screens. Remember,

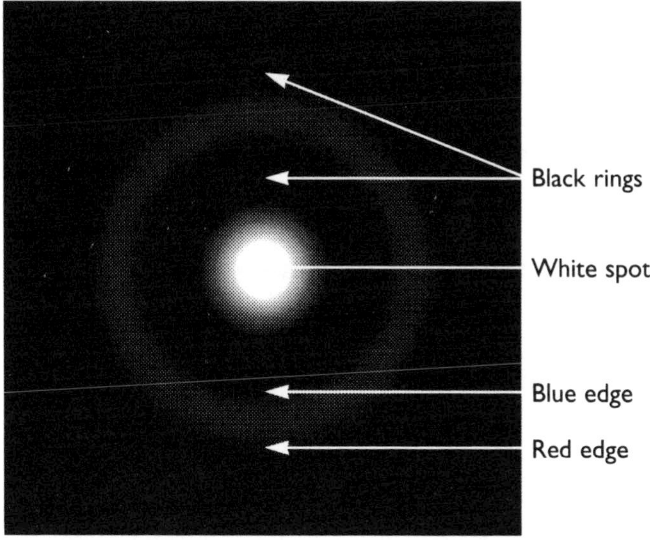

FIGURE 2.5 *The pattern of light and shadow formed by white light after passing through a small circular hole.*

there is nothing happening here but the casting of a shadow. Figure 2.5 is just the shadow that would be cast by the second screen in Figure 2.4. If light travelled only in straight lines, there would only be a tiny white dot (much smaller than the central bright spot in Figure 2.5), surrounded by a very narrow penumbra. Outside that there would be pure umbra – total darkness.

Puzzling though it may be that light rays should bend when passing through small holes, it is not, I think, fundamentally disturbing. In any case, what matters for our present purposes is that it does bend. This means that shadows in general need not look like silhouettes of the objects that cast them. What is more, this is not just a matter of blurring, caused by penumbras. It turns out that an obstacle with an intricate pattern of holes can cast a shadow of an entirely different pattern.

Figure 2.6 shows, at roughly its actual size, a part of the pattern of shadows cast three metres from a pair of straight, parallel slits in an otherwise opaque barrier. The slits are one-fifth of a

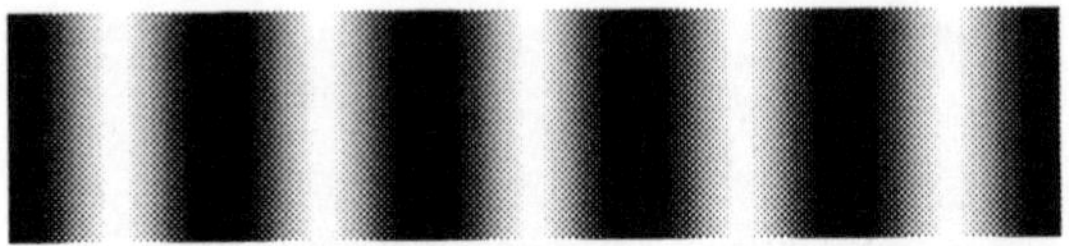

FIGURE 2.6 *The shadow cast by a barrier containing two straight, parallel slits.*

millimetre apart, and illuminated by a parallel-sided beam of pure red light from a laser on the other side of the barrier. Why laser light and not torchlight? Only because the precise shape of a shadow also depends on the colour of the light in which it is cast; white light, as produced by a torch, contains a mixture of all visible colours, so it can cast shadows with multicoloured fringes. Therefore in experiments about the precise shapes of shadows we are better off using light of a single colour. We could put a coloured filter (such as a pane of coloured glass) over the front of the torch, so that only light of that colour would get through. That would help, but filters are not all that discriminating. A better method is to use laser light, for lasers can be tuned very accurately to emit light of whatever colour we choose, with almost no other colour present.

If light travelled in straight lines, the pattern in Figure 2.6 would consist simply of a pair of bright bands one-fifth of a millimetre apart (too close to distinguish on this scale), with sharp edges and with the rest of the screen in shadow. But in reality the light bends in such a way as to make many bright bands and dark bands, and no sharp edges at all. If the slits are moved sideways, so long as they remain within the laser beam, the pattern also moves by the same amount. In this respect it behaves exactly like an ordinary large-scale shadow. Now, what sort of shadow is cast if we cut a second, identical pair of slits in the barrier, interleaved with the existing pair, so that we have four slits at intervals of one-tenth of a millimetre? We might expect the pattern to look almost exactly like Figure 2.6. After all, the first pair of slits, by itself, casts the shadows in Figure 2.6, and as I have just said, the second pair, by itself, would cast the same pattern, shifted about a tenth of a millimetre to the side – in almost the same place. We even know that light beams normally pass through each other unaffected. So the two pairs of slits together should give essentially the same pattern again, though twice as bright and slightly more blurred.

In reality, though, what happens is nothing like that. The real shadow of a barrier with four straight, parallel slits is shown in Figure 2.7(a). For comparison I have repeated, below it, the illustration of the two-slit pattern (Figure 2.7(b)). Clearly, the four-slit shadow is not a combination of two slightly displaced two-slit shadows, but has a new and more complicated pattern. In this pattern there are places, such as the point marked X, which are dark on the four-slit pattern, but bright on the two-slit pattern. These places were bright when there were two slits in the barrier, but *went dark* when we cut a second pair of slits for the light to pass through. Opening those slits has *interfered* with the light that was previously arriving at X.

So, adding two more light sources darkens the point X; removing them illuminates it again. How? One might imagine two photons heading towards X and bouncing off each other like billiard balls. Either photon alone would have hit X, but the two together interfere with each other so that they both end up elsewhere. I shall show in a moment that this explanation cannot be true. Nevertheless, the basic idea of it is inescapable: *something* must be coming through that second pair of slits to prevent the light from the first pair from

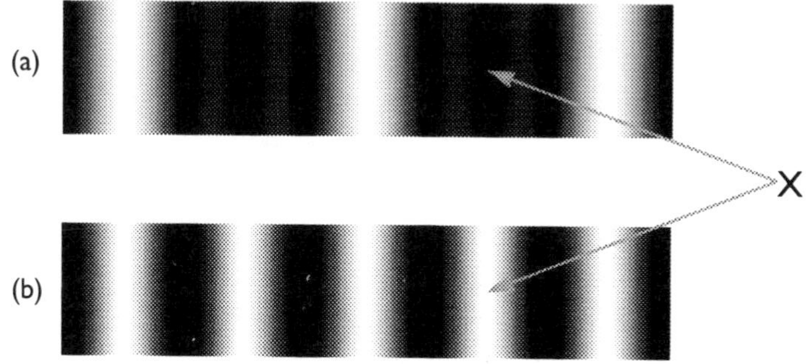

FIGURE 2.7 *The shadows cast by a barrier containing (a) four and (b) two straight, parallel slits.*

reaching X. But what? We can find out with the help of some further experiments.

First, the four-slit pattern of Figure 2.7(a) appears only if all four slits are illuminated by the laser beam. If only two of them are illuminated, a two-slit pattern appears. If three are illuminated, a three-slit pattern appears, which looks different again. So whatever causes the interference is in the light beam. The two-slit pattern also reappears if two of the slits are filled by anything opaque, but not if they are filled by anything transparent. In other words, the interfering entity is obstructed by anything that obstructs light, even something as insubstantial as fog. But it can penetrate anything that allows light to pass, even something as impenetrable (to matter) as diamond. If complicated systems of mirrors and lenses are placed anywhere in the apparatus, so long as light can travel from each slit to a particular point on the screen, what will be observed at that point will be part of a four-slit pattern. If light from only two slits can reach a particular point, part of a two-slit pattern will be observed there, and so on.

So, whatever causes interference behaves like light. It is found everywhere in the light beam and nowhere outside it. It is reflected, transmitted or blocked by whatever reflects, transmits or blocks light. You may be wondering why I am labouring this point. Surely it is obvious that it *is* light; that is, what interferes with photons from each slit is photons from the other slits. But you may be inclined to doubt the obvious after the next experiment, the denouement of the series.

What should we expect to happen when these experiments are performed *with only one photon at a time*? For instance, suppose that our torch is moved so far away that only one photon per day is falling on the screen. What will our frog, observing from the screen, see? If it is true that what interferes with each photon is other photons, then shouldn't the interference be lessened when the photons are very sparse? Should it not cease altogether when there is only one photon passing through the apparatus at any one time? We might still expect penumbras, since a photon might be capable of changing course when passing through a slit (perhaps by striking a glancing blow at the edge). But what we surely could not observe is any place on the screen, such as X, that receives photons when two slits are open, but which *goes dark* when two more are opened.

Yet that is exactly what we do observe. However sparse the photons are, the shadow pattern remains the same. Even when the experiment is done with one photon at a time, none of them is ever observed to arrive at X when all four slits are open. Yet we need only close two slits for the flickering at X to resume.

Could it be that the photon splits into fragments which, after passing through the slits, change course and recombine? We can rule that possibility out too. If, again, we fire one photon through the apparatus, but use four detectors, one at each slit, then at most one of them ever registers anything. Since in such an experiment we never observe two of the detecters going off at once, we can tell that the entities that they detect are not splitting up.

So, if the photons do not split into fragments, and are not being deflected by other photons, what does deflect them? When a single photon at a time is passing through the apparatus, what can be coming through the other slits to interfere with it?

Let us take stock. We have found that when one photon passes through this apparatus,

> it passes through one of the slits, and then something interferes with it, deflecting it in a way that depends on what other slits are open;
>
> the interfering entities have passed through some of the other slits;
>
> the interfering entities behave exactly like photons . . .
>
> . . . except that they cannot be seen.

I shall now start calling the interfering entities 'photons'. That is what they are, though for the moment it does appear that photons come in two sorts, which I shall temporarily call *tangible* photons and *shadow* photons. Tangible photons are the ones we can see, or detect with instruments, whereas the shadow photons are intangible (invisible) – detectable only indirectly through their interference effects on the tangible photons. (Later, we shall see that there is no intrinsic difference between tangible and shadow photons: each photon is tangible in one universe and intangible in all the other parallel universes – but I anticipate.) What we have inferred so far is only that each tangible photon has an accompanying retinue of shadow photons, and that when a photon passes through one of our four slits, some shadow photons pass through the other three slits. Since different interference patterns appear when we cut slits at other places in the screen, provided that they are within the beam, shadow photons must be arriving all over the illuminated part of the screen whenever a tangible photon arrives. Therefore there are many more shadow photons than tangible ones. How many? Experiments cannot put an upper bound on the number, but they do set a rough lower bound. In a laboratory the largest area that we could conveniently illuminate with a laser might be about a square metre, and the smallest manageable size for the holes might be about a thousandth of a millimetre. So there are about 10^{12} (one trillion) possible hole-locations on the screen. Therefore there must be at least a trillion shadow photons accompanying each tangible one.

Thus we have inferred the existence of a seething, prodigiously complicated, hidden world of shadow photons. They travel at the speed of light, bounce off mirrors, are refracted by lenses, and are stopped by opaque barriers or filters of the wrong colour. Yet they do not trigger even the most sensitive detectors. The only thing in the universe that a shadow photon can be observed to affect is the tangible photon that it accompanies. That is the phenomenon of interference. Shadow photons would go entirely unnoticed were it not for this phenomenon and the strange patterns of shadows by which we observe it.

Interference is not a special property of photons alone. Quantum theory predicts, and experiment confirms, that it occurs for every sort of particle. So there must be hosts of shadow neutrons accompanying every tangible neutron, hosts of shadow electrons accompanying every electron, and so on. Each of these shadow particles is detectable only indirectly, through its interference with the motion of its tangible counterpart.

It follows that reality is a much bigger thing than it seems, and most of it is invisible. The objects and events that we and our instruments can directly observe are the merest tip of the iceberg.

Now, tangible particles have a property that entitles us to call them, collectively, a *universe*. This is simply their defining property of being tangible, that is, of interacting with each other, and hence of being directly detectable by instruments and sense organs made of other tangible particles. Because of the phenomenon of interference, they are not *wholly* partitioned off from the rest of reality (that is, from the shadow particles). If they were, we should never have discovered that there is more to reality than tangible particles. But to a good approximation they do resemble the universe that we see around us in everyday life, and the universe referred to in classical (pre-quantum) physics.

For similar reasons, we might think of calling the shadow particles, collectively, a *parallel universe*, for they too are affected by tangible particles only through interference phenomena. But we can do better than that. For it turns out that shadow particles are partitioned among themselves in exactly the same way as the universe of tangible particles is partitioned from them. In other words, they do not form a single, homogeneous parallel universe vastly larger than the tangible one, but rather a huge number of parallel universes, each similar in composition to the tangible one, and each obeying the same laws of physics, but differing in that the particles are in different positions in each universe.

A remark about terminology. The word 'universe' has traditionally been used to mean 'the whole of physical reality'. In that sense there can be at most one universe. We could stick to that definition, and say that the entity we have been accustomed to calling 'the universe' – namely, all the directly perceptible matter and energy around us, and the surrounding space – is not the whole universe after all, but only a small portion of it. Then we should have to invent a new name for that small, tangible portion. But most physicists prefer to carry on using the word 'universe' to denote the same entity that it has always denoted, even though that entity now turns out to be only a small part of physical reality. A new word, *multiverse*, has been coined to denote physical reality as a whole.

Single-particle interference experiments such as I have been describing show us that the multiverse exists and that it contains many counterparts of each particle in the tangible universe. To reach the further conclusion that the multiverse is roughly partitioned into parallel universes, we must consider interference phenomena involving more than one tangible particle. The simplest way of doing this is to ask, by way of a 'thought experiment', what must be happening at the microscopic level when shadow photons strike an opaque object. They are stopped, of course: we know that because interference ceases when an opaque barrier is placed in the paths of shadow photons. But why? What stops them? We can rule out the straightforward answer – that they are absorbed, like tangible photons would be, by the tangible atoms in the barrier. For one thing, we know that shadow photons do not interact with tangible atoms. For another, we can verify by measuring the atoms in the barrier (or more precisely, by replacing the barrier by a detector) that they neither absorb energy nor change their state in any way unless they are struck by tangible photons. Shadow photons have no effect.

To put that another way, shadow photons and tangible photons are affected in identical ways when they reach a given barrier, but the barrier itself is not identically affected by the two types of photon. In fact, as far as we can tell, it is not affected by shadow photons at all. That is indeed the defining property of shadow photons, for if any material were observably affected by them, that material could be used as a shadow-photon detector and the entire phenomenon of shadows and interference would not be as I have described it.

Hence there is some sort of shadow barrier at the same location as the tangible barrier. It takes no great leap of imagination to conclude that this shadow barrier is made up of the *shadow atoms* that we already know must be present as counterparts of the tangible atoms in the barrier. There are very many of them present for each tangible atom. Indeed, the total density of shadow atoms in even the lightest fog would be more than sufficient to stop a tank, let alone a photon, *if* they could all affect it. Since we find that partially transparent barriers have the same degree of transparency for shadow photons as for tangible ones, it follows that not all the shadow atoms in the path of a particular shadow photon can be involved in blocking its passage. Each shadow photon encounters much the same sort of barrier as its tangible counterpart does, a barrier consisting of only a tiny proportion of all the shadow atoms that are present.

For the same reason, each shadow atom in the barrier can be interacting with only a small proportion of the other shadow atoms in its vicinity, and the ones it does interact with form a barrier much like the tangible one. And so on. All matter, and all physical processes, have this structure. If the tangible barrier is the frog's retina, then there must be many shadow retinas, each capable of stopping only one of the shadow-counterparts of each photon. Each shadow retina only interacts strongly with the corresponding shadow photons, and with the corresponding shadow frog, and so on. In other words, particles are grouped into parallel universes. They are 'parallel' in the sense that within each universe particles interact with each other just as they do in the tangible universe, but each universe affects the others only weakly, through interference phenomena.

Thus we have reached the conclusion of the chain of reasoning that begins with strangely shaped shadows and ends with parallel universes. Each step takes the form of noting that the behaviour of objects that we observe can be explained only if there are unobserved objects present, and if those unobserved objects have certain properties. The heart of the argument is that single-particle interference phenomena unequivocally rule out the possibility that the tangible universe around us is all that exists. There is no disputing the fact that such interference phenomena occur. Yet the existence of the multiverse is still a minority view among physicists. Why?

The answer, I regret to say, does not reflect well upon the

majority. I shall have more to say about this in Chapter 13, but for the moment let me point out that the arguments I have presented in this chapter are compelling only to those who seek explanations. Those who are satisfied with mere prediction, and who have no strong desire to understand how the predicted outcomes of experiments come about, may if they wish simply deny the existence of anything other than what I have been calling 'tangible' entities. Some people, such as instrumentalists and positivists, take this line as a matter of philosophical principle. I have already said what I think of such principles, and why. Other people just don't want to think about it. After all, it is such a *large* conclusion, and such a disturbing one on first hearing. But I think that those people are making a mistake. As I hope to persuade readers who bear with me, understanding the multiverse is a precondition for understanding reality as best we can. Nor is this said in a spirit of grim determination to seek the truth no matter how unpalatable it may be (though I hope I would take that attitude if it came to it). It is, on the contrary, because the resulting world-view is so much more integrated, and makes more sense in so many ways, than any previous world-view, and certainly more than the cynical pragmatism which too often nowadays serves as a surrogate for a world-view among scientists.

'Why can't we just say,' some pragmatic physicists ask, 'that photons behave *as if* they were interacting with invisible entities? Why can we not leave it at that? Why do we have to go on to take a position about whether those invisible entities are really there?' A more exotic variant of what is essentially the same idea is the following. 'A tangible photon is real; a shadow photon is merely a way in which the real photon could possibly have behaved, but did not. Thus quantum theory is about the *interaction of the real with the possible*.' This, at least, sounds suitably profound. But unfortunately the people who take either of these views – including some eminent scientists who ought to know better – invariably lapse into mumbo-jumbo at that point. So let us keep cool heads. The key fact is that a real, tangible photon *behaves differently* according to what paths are open, elsewhere in the apparatus, for something to travel along and eventually intercept the tangible photon. Something does travel along those paths, and to refuse to call it 'real' is merely to play with words. 'The possible' cannot interact with the real: non-existent entities cannot deflect real ones from their paths. If a photon is deflected, it must have been deflected by something, and I have called that thing a 'shadow photon'. Giving it a name does not make it real, but it cannot be true that an actual event, such as the arrival and detection of a tangible photon, is caused by an imaginary event such as what that photon 'could have done' but did not do. It is only what really happens that can cause other things really to happen. If the complex motions of the shadow photons in an interference experiment were mere possibilities that did not in fact take place, then the interference phenomena we see would not, in fact, take place.

The reason why interference effects are usually so weak and hard to detect can be found in the quantum-mechanical laws that govern them. Two particular implications of those laws are relevant. First, every subatomic particle has counterparts in other universes, and is interfered with only by those counterparts. It is not directly affected by any other particles in those universes. Therefore interference is observed only in special situations where the paths of a particle and its shadow counterparts separate and then reconverge (as when a photon and shadow photon are heading towards the same point on the screen). Even the timing must be right: if one of the two paths involves a delay, the interference is reduced or prevented. Second, the detection of interference between any two universes requires an interaction to take place between *all the particles whose positions and other attributes are not identical* in the two universes. In practice this means that interference is strong enough to be detected only between universes that are very alike. For example, in all the experiments I have described, the interfering universes differ only in the position of one photon. If a photon affects other particles in its travels, and in particular if it is observed, then those particles or the observer will also become differentiated in different universes. If so, subsequent interference involving that photon will be undetectable in practice because the requisite interaction between *all* the affected particles is too complicated to arrange. I must mention here that the standard phrase for describing this fact, namely 'observation destroys interference', is very misleading in three ways. First, it suggests some sort of psychokinetic effect of the conscious 'observer' on basic physical phenomena, though there is no such effect. Second, the interference is not 'destroyed': it is just (much!) harder to observe because doing so involves controlling the precise behaviour of many more particles. And third, it is not just 'observation', but *any* effect of the photon on its surroundings that depends on which path the photon has taken, that does this.

For the benefit of readers who may have seen other accounts of quantum physics, I must briefly make contact between the argument I have given in this chapter and the way the subject is usually presented. Perhaps because the debate began among theoretical physicists, the traditional starting-point has been quantum theory itself. One states the theory as carefully as possible, and then one tries to understand what it tells us about reality. That is the only possible approach if one wants to understand the finer details of quantum phenomena. But as regards the issue of whether reality consists of one universe or many, it is an unnecessarily complicated approach. That is why I have not followed it in this chapter. I have not even stated any of the postulates of quantum theory – I have merely described some physical phenomena and drawn inescapable conclusions. But if one does start from theory, there are two things that everyone agrees on. The first is that quantum theory is unrivalled in its ability to predict the outcomes of experiments, even if one blindly uses its equations without worrying much about what they mean. The second is that quantum theory tells us something new and bizarre about the nature of reality. The dispute is only about what exactly this is. The physicist Hugh Everett was the first to understand clearly (in 1957, some thirty years after the theory became the basis of subatomic physics) that quantum theory describes a multiverse. Ever since, the argument has raged about whether the theory admits of any other interpretation (or re-interpretation, or reformulation, or modification, etc.) in which it describes a single universe, but continues correctly to predict the outcomes of experiments. In other words, does accepting the predictions of quantum theory force us to accept the existence of parallel universes?

It seems to me that this question, and therefore the whole prevailing tone of the debate on this issue, is wrong-headed. Admittedly, it is right and proper for theoretical physicists such as myself to devote a great deal of effort to trying to understand the formal structure of quantum theory, but not at the expense of losing sight of our primary objective, which is to understand reality. Even if the predictions of quantum theory could, somehow, be made without referring to more than one universe, individual photons would still cast shadows in the way I have described. Without knowing anything of quantum theory, one can see that those shadows could not be the result of any single history of the photon as it travels from the torch to the observer's eye. They are incompatible with any explanation in terms of only the photons that we see. Or in terms of only the barrier that we see. Or in terms of only the universe that we see. Therefore, if the best theory available to physics did not refer to parallel universes, it would merely mean that we needed a better theory, one that did refer to parallel universes, in order to explain what we see.

So, does accepting the predictions of quantum theory force us to accept the existence of parallel universes? Not in itself. We can always reinterpret any theory along instrumentalist lines so that it does not force us to accept anything about reality. But that is beside the point. As I have just said, we do not need deep theories to tell us that parallel universes exist – single-particle interference phenomena tell us that. What we need deep theories for is to explain and predict such phenomena: to tell us what the other universes are like, what laws they obey, how they affect one another, and how all this fits in with the theoretical foundations of other subjects. That is what quantum theory does. The quantum theory of parallel universes is not the problem, it is the solution. It is not some troublesome, optional interpretation emerging from arcane theoretical considerations. It is the explanation – the only one that is tenable – of a remarkable and counter-intuitive reality.

So far, I have been using temporary terminology which suggests that one of the many parallel universes differs from the others by being 'tangible'. It is time to sever that last link with the classical, single-universe conception of reality. Let us go back to our frog. We have seen that the story of the frog that stares at the distant torch for days at a time, waiting for the flicker that comes on average once a day, is not the whole story, because there must also be shadow frogs, in shadow universes that co-exist with the tangible one, also waiting for photons. Suppose that our frog is trained to jump when it sees a flicker. At the beginning of the experiment, the tangible frog will have a large set of shadow counterparts, all initially alike. But shortly afterwards they will no longer all be alike. Any particular one of them is unlikely to see a photon immediately. But what is a rare event in any one universe is a common event in the multiverse as a whole. At any instant, somewhere in the multiverse, there are a few universes in which one of the photons is currently striking the retina of the frog in that universe. And that frog jumps.

Why exactly does it jump? Because within its universe it obeys the same laws of physics as tangible frogs do, and its shadow retina has been struck by a shadow photon belonging to that universe. One of the light-sensitive shadow molecules in that shadow retina has responded by undergoing complex chemical changes, to which the shadow frog's optic nerve has in turn responded. It has transmitted a message to the shadow frog's brain, and the frog has consequently experienced the sensation of seeing a flicker.

Or should I say 'the *shadow* sensation of seeing a flicker'? Surely not. If 'shadow' observers, be they frogs or people, are real, then their sensations must be real too. When they observe what we might call a shadow object, they observe that it is tangible. They observe this by the same means, and according to the same definition, as we apply when we say that the universe we observe is 'tangible'. Tangibility is relative to a given observer. So objectively there are not two kinds of photon, tangible and shadow, nor two kinds of frog, nor two kinds of universe, one tangible and the rest shadow. There is nothing in the description I have given of the formation of shadows, or any of the related phenomena, that distinguishes between the 'tangible' and the 'shadow' objects, apart from the mere assertion that one of the copies is 'tangible'. When I introduced tangible and shadow photons I apparently distinguished them by saying that we can see the former, but not the latter. But who are 'we'? While I was writing that, hosts of shadow Davids were writing it too. They too drew a distinction between tangible and shadow photons; but the photons they called 'shadow' include the ones I called 'tangible', and the photons they called 'tangible' are among those I called 'shadow'.

Not only do none of the copies of an object have any privileged position in the explanation of shadows that I have just outlined, neither do they have a privileged position in the full mathematical explanation provided by quantum theory. I may feel subjectively that I am distinguished among the copies as the 'tangible' one, because I can directly perceive myself and not the others, but I must come to terms with the fact that all the others feel the same about themselves.

Many of those Davids are at this moment writing these very words. Some are putting it better. Others have gone for a cup of tea.

TERMINOLOGY

photon A particle of light.

tangible/shadow For the purposes of exposition in this chapter only, I called particles in this universe *tangible*, and particles in other universes *shadow particles*.

multiverse The whole of physical reality. It contains many parallel universes.

parallel universes They are 'parallel' in the sense that within each universe particles interact with each other just as they do in the tangible universe, but each universe affects the others only weakly, through interference phenomena.

quantum theory The theory of the physics of the multiverse.

quantization The property of having a discrete (rather than continuous) set of possible values. Quantum theory gets its name from its assertion that all measurable quantities are quantized. However, the most significant quantum effect is not quantization but interference.

interference The effect of a particle in one universe on its counterpart in another. Photon interference can cause shadows to be much more complicated than mere silhouettes of the obstacles causing them.

SUMMARY

In interference experiments there can be places in a shadow-pattern that go dark when new openings are made in the barrier casting the shadow. This remains true even when the experiment is performed with individual particles. A chain of reasoning based on this fact rules out the possibility that the universe we see around us constitutes the whole of reality. In fact the whole of physical reality, the multiverse, contains vast numbers of parallel universes.

Quantum physics is one of the four main strands of explanation. The next strand is epistemology, *the theory of knowledge.*

3

Problem-solving

I do not know which is stranger – the behaviour of shadows itself, or the fact that contemplating a few patterns of light and shadow can force us to revise so radically our conception of the structure of reality. The argument I have outlined in the previous chapter is, notwithstanding its controversial conclusion, a typical piece of scientific reasoning. It is worth reflecting on the character of this reasoning, which is itself a natural phenomenon at least as surprising and full of ramifications as the physics of shadows.

To those who would prefer reality to have a more prosaic structure, it may seem somehow out of proportion – unfair, even – that such momentous consequences can flow from the fact that a tiny spot of light on a screen should be *here* rather than *there*. Yet they do, and this is by no means the first time in the history of science that such a thing has happened. In this respect the discovery of other universes is quite reminiscent of the discovery of other planets by early astronomers. Before we sent space probes to the Moon and planets, *all* our information about planets came from spots of light (or other radiation) being observed in one place rather than another. Consider how the original, defining fact about planets – the fact that they are not stars – was discovered. Watching the night sky for a few hours, one sees that the stars appear to revolve about a particular point in the sky. They revolve rigidly, holding fixed positions relative to one another. The traditional explanation was that the night sky was a huge 'celestial sphere' revolving around the fixed Earth, and that the stars were either holes in the sphere or glowing embedded crystals. However, among the thousands of points of light in the sky visible to the naked eye, there are a handful of the brightest which, over longer periods, do not move as if they were fixed on a celestial sphere. They wander about the sky in more complex motions. They are called 'planets', from the Greek word meaning 'wanderer'. Their wandering was a sign that the celestial-sphere explanation was inadequate.

Successive explanations of the motions of planets have played an important role in the history of science. Copernicus's *heliocentric theory* placed the planets and the Earth in circular orbits round the Sun. Kepler discovered that the orbits are ellipses rather than circles. Newton explained the ellipses through his inverse-square law of gravitational forces, and his theory was later used to predict that the mutual gravitational attraction of planets would cause small deviations from elliptical orbits. The observation of such deviations led to the discovery in 1846 of a new planet, Neptune, one of many discoveries that spectacularly corroborated Newton's theory. Nevertheless, a few decades later Einstein's general theory of relativity gave us a fundamentally different explanation of gravity, in terms of curved space and time, and thereby predicted slightly different motions again. For instance, it correctly predicted that every year the planet Mercury would drift by about one ten-thousandth of a degree away from where Newton's theory said it should be. It also implied that starlight passing close to the Sun would be deflected twice as much by gravity as Newton's theory would predict. The observation of this deflection by Arthur Eddington in 1919 is often deemed to mark the moment at which the Newtonian world-view ceased to be rationally tenable. (Ironically, modern reappraisals of the accuracy of Eddington's experiment suggest that this may have been premature.) The experiment, which has since been repeated with great accuracy, involved measuring the positions of spots (the images of stars close to the limb of the Sun during an eclipse) on a photographic plate.

As astronomical predictions became more accurate, the differences between what successive theories predicted about the appearance of the night sky diminished. Ever more powerful telescopes and measuring instruments have had to be constructed to detect the differences. However, the explanations underlying these predictions have not been converging. On the contrary, as I have just outlined, there has been a succession of revolutionary changes. Thus observations of ever smaller physical effects have been forcing ever greater changes in our world-view. It may therefore seem that we are inferring ever grander conclusions from ever scantier evidence. What justifies these inferences? Can we be sure that just because a star appeared millimetrically displaced on Eddington's photographic plate, space and time must be curved; or that because a photodetector at a certain position does not register a 'hit' in weak light, there must be parallel universes?

Indeed, what I have just said understates both the fragility and the indirectness of all experimental evidence. For we do not directly perceive the stars, spots on photographic plates, or any other external objects or events. We see things only when images of them appear on our retinas, and we do not perceive even those images until they have given rise to electrical impulses in our nerves, and those impulses have been received and interpreted by our brains. Thus the physical evidence that directly sways us, and causes us to adopt one theory or world-view rather than another, is less than millimetric: it is measured in thousandths of a millimetre (the separation of nerve fibres in the optic nerve), and in hundredths of a volt (the change in electric potential in our nerves that makes the difference between our perceiving one thing and perceiving another).

However, we do not accord equal significance to all our sensory impressions. In scientific experiments we go to great lengths to bring to our perceptions those aspects of external reality that we think might help us to distinguish between rival theories we are considering. Before we even make an observation, we decide carefully where and when we should look, and what we should look for. Often we use complex, specially constructed instruments, such as telescopes and photomultipliers. Yet however sophisticated the instruments we use, and however substantial the external causes to which we attribute their readings, we perceive those readings exclusively through our own sense organs. There is no getting away from the fact that we human beings are small creatures with only a few inaccurate, incomplete channels through which we receive all information from outside ourselves. We interpret this information as evidence of a large and complex external universe (or multiverse). But when we are weighing up this evidence, we are literally contemplating nothing more than patterns of weak electric current trickling through our own brains.

What justifies the inferences we draw from these patterns? It is certainly not a matter of logical deduction. There is no way of *proving* from these or from any other observations that the external universe, or multiverse, exists at all, let alone that the electric currents received by our brains stand in any particular relationship to it. Anything or everything that we perceive might be an illusion or a dream. Illusions and dreams are, after all, common. *Solipsism*, the theory that only one mind exists and that what appears to be external reality is only a dream taking place in that mind, cannot be logically disproved. Reality *might* consist of one person, presumably you, dreaming a lifetime's experiences. Or it might consist of just you and me. Or just the planet Earth and its inhabitants. And if we dreamed evidence – any evidence – of the existence of other people, or other planets, or other universes, that would prove nothing about how many of those things there really are.

Since solipsism, and an infinity of related theories, are logically consistent with your perceiving any possible observational evidence, it follows that you can logically deduce nothing about reality from observational evidence. How, then, could I say that the observed behaviour of shadows 'rules out' the theory that there is only one universe, or that eclipse observations make the Newtonian world-view 'rationally untenable'? How can that be so? If 'ruling out' does not mean 'disproving', what does it mean? Why should we feel compelled to change our world-view, or indeed any opinion at all, on account of something being 'ruled out' in that sense? This critique seems to cast doubt on the whole of science – on any reasoning about external reality that appeals to observational evidence. If scientific reasoning does not amount to sequences of logical deductions from the evidence, what does it amount to? Why should we accept its conclusions?

This is known as the 'problem of induction'. The name derives from what was, for most of the history of science, the prevailing theory of how science works. The theory was that there exists, short of mathematical proof, a lesser but still worthy form of justification called *induction*. Induction was contrasted, on the one hand, with the supposedly perfect justification provided by *de*duction, and on the other hand with supposedly weaker philosophical or intuitive forms of reasoning that do not even have observational evidence to back them up. In the inductivist theory of scientific knowledge, observations play two roles: first, in the discovery of scientific theories, and second, in their justification. A theory is supposed to be discovered by 'extrapolating' or 'generalizing' the results of observations. Then, if large numbers of observations conform to the theory, and none deviates from it, the theory is supposed to be justified – made more believable, probable or reliable. The scheme is illustrated in Figure 3.1.

The inductivist analysis of my discussion of shadows would therefore go something like this: 'We make a series of observations of shadows, and see interference phenomena (stage 1). The results conform to what would be expected if there existed parallel universes which affect one another in certain ways. But at first no one notices this. Eventually (stage 2) someone forms the generalization that interference will *always* be observed under the given circumstances, and thereby induces the theory that parallel universes are responsible. With every further observation of interference (stage 3) we become a little more convinced of that theory. After a sufficiently long sequence of such observations, and provided that none of them ever contradicts the theory, we conclude (stage 4) that the theory is true. Although we can never be absolutely sure, we are for practical purposes convinced.'

It is hard to know where to begin in criticizing the inductivist

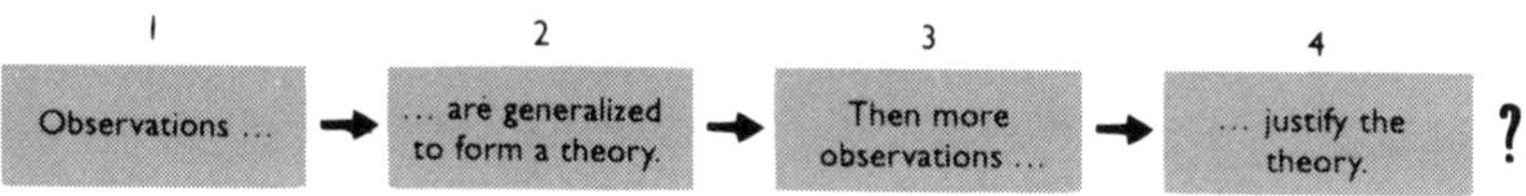

FIGURE 3.1 *The inductivist scheme.*

conception of science – it is so profoundly false in so many different ways. Perhaps the worst flaw, from my point of view, is the sheer *non sequitur* that a generalized prediction is tantamount to a new theory. Like all scientific theories of any depth, the theory that there are parallel universes simply does not have the form of a generalization from the observations. Did we observe first one universe, then a second and a third, and then induce that there are trillions of them? Was the generalization that planets will 'wander' round the sky in one pattern rather than another, equivalent to the theory that planets are worlds, in orbit round the Sun, and that the Earth is one of them? It is also not true that repeating our observations is the way in which we become convinced of scientific theories. As I have said, theories are explanations, not merely predictions. If one does not accept a proposed explanation of a set of observations, making the observations over and over again is seldom the remedy. Still less can it help us to create a satisfactory explanation when we cannot think of one at all.

Furthermore, even mere predictions can never be justified by observational evidence, as Bertrand Russell illustrated in his story of the chicken. (To avoid any possible misunderstanding, let me stress that this was a metaphorical, anthropomorphic chicken, representing a human being trying to understand the regularities of the universe.) The chicken noticed that the farmer came every day to feed it. It predicted that the farmer would continue to bring food every day. Inductivists think that the chicken had 'extrapolated' its observations into a theory, and that each feeding time added justification to that theory. Then one day the farmer came and wrung the chicken's neck. The disappointment experienced by Russell's chicken has also been experienced by trillions of other chickens. This inductively justifies the conclusion that induction cannot justify any conclusions!

However, this line of criticism lets inductivism off far too lightly. It does illustrate the fact that repeated observations cannot *justify* theories, but in doing so it entirely misses (or rather, accepts) a more basic misconception: namely, that the inductive extrapolation of observations to *form* new theories is even possible. In fact, it is impossible to extrapolate observations unless one has already placed them within an explanatory framework. For example, in order to 'induce' its false prediction, Russell's chicken must first have had in mind a false explanation of the farmer's behaviour. Perhaps it guessed that the farmer harboured benevolent feelings towards chickens. Had it guessed a different explanation – that the farmer was trying to fatten the chickens up for slaughter, for instance – it would have 'extrapolated' the behaviour differently. Suppose that one day the farmer starts bringing the chickens more food than usual. How one extrapolates this new set of observations to predict the farmer's future behaviour depends entirely on how one explains it. According to the benevolent-farmer theory, it is evidence that the farmer's benevolence towards chickens has increased, and that therefore the chickens have even less to worry about than before. But according to the fattening-up theory, the behaviour is ominous – it is evidence that slaughter is imminent.

The fact that the same observational evidence can be 'extrapolated' to give two diametrically opposite predictions according to which explanation one adopts, and cannot justify either of them, is not some accidental limitation of the farmyard environment: it is true of all observational evidence under all circumstances. Observations could not possibly play either of the roles assigned to them in the inductivist scheme, even in respect of mere predictions, let alone genuine explanatory theories. Admittedly, inductivism is based on the common-sense theory of the growth of knowledge – that we learn from experience – and historically it was associated with the liberation of science from dogma and tyranny. But if we want to understand the true nature of knowledge, and its place in the fabric of reality, we must face up to the fact that inductivism is false, root and branch. No scientific reasoning, and indeed no successful reasoning of any kind, has ever fitted the inductivist description.

What, then, *is* the pattern of scientific reasoning and discovery? We have seen that inductivism and all other prediction-centred theories of knowledge are based on a misconception. What we need is an explanation-centred theory of knowledge: a theory of how explanations come into being and how they are justified; a theory of how, why and when we should allow our perceptions to change our world-view. Once we have such a theory, we need no separate theory of predictions. For, given an explanation of some observable phenomenon, it is no mystery how one obtains predictions. And if one has justified an explanation, then any predictions derived from that explanation are automatically justified too.

Fortunately, the prevailing theory of scientific knowledge, which in its modern form is due largely to the philosopher Karl Popper (and which is one of my four 'main strands' of explanation of the fabric of reality), can indeed be regarded as a theory of explanations in this sense. It regards science as a *problem-solving* process. Inductivism regards the catalogue of our past observations as a sort of skeletal theory, supposing that science is all about filling in the gaps in that theory by interpolation and extrapolation. Problem-solving does begin with an inadequate theory – but not with the notional 'theory' consisting of past observations. It begins with our best existing theories. When some of those theories seem inadequate to us, and we want new ones, that is what constitutes a *problem*. Thus, contrary to the inductivist scheme shown in Figure 3.1, scientific discovery need not begin with observational evidence. But it does always begin with a problem. By a 'problem' I do not necessarily mean a practical emergency, or a source of anxiety. I just mean a set of ideas that seems inadequate and worth trying to improve. The existing explanation may seem too glib, or too laboured; it may seem unnecessarily narrow, or unrealistically ambitious. One may glimpse a possible unification with other ideas. Or a satisfactory explanation in one field may appear to be irreconcilable with an equally satisfactory explanation in another. Or it *may* be that there have been some surprising observations – such as the wandering of planets – which existing theories did not predict and cannot explain.

This last type of problem resembles stage 1 of the inductivist scheme, but only superficially. For an unexpected observation never initiates a scientific discovery unless the pre-existing theories already contain the seeds of the problem. For example, clouds wander even more than planets do. This unpredictable wandering was presumably familiar long before planets were discovered. Moreover, predicting the weather would always have been valuable to farmers, seafarers and soldiers, so there would always have been an incentive to theorize about how clouds move. Yet it was not meteorology that blazed the trail for modern science, but astronomy. Observational evidence about meteorology was far more readily available than in astronomy, but no one paid much attention to it, and no one induced any theories from it about cold fronts or anticyclones. The history of science was not crowded with disputes, dogmas, heresies, speculations and elaborate theories about the nature of clouds and their motion. Why? Because under the established explanatory structure for weather, it was perfectly comprehensible that cloud motion should be unpredictable. Common sense suggests that clouds move with the wind. When they drift in other directions, it is reasonable to surmise that the wind can be different at different altitudes, and is rather unpredictable, and so it is easy to conclude that there is no more to be explained. Some people, no doubt, took this view about planets, and assumed that they were just glowing objects on the celestial sphere, blown about by high-altitude winds, or perhaps moved by angels, and that there was no more to be explained. But others were not satisfied with that, and guessed that there were deeper explanations behind the wandering of planets. So they searched for such explanations, and found them. At various times in the history of astronomy there appeared to be a mass of unexplained observational evidence; at other times only a scintilla, or none at all. But always, if people had chosen what to theorize about according to the cumulative number of observations of particular phenomena, they would have chosen clouds rather than planets. Yet they chose planets, and for diverse reasons. Some reasons depended on preconceptions about how cosmology ought to be, or on arguments advanced by ancient philosophers, or on mystical numerology. Some were based on the physics of the day, others on mathematics or geometry. Some have turned out to have objective merit, others not. But every one of them amounted to this: it seemed to someone that the existing explanations could and should be improved upon.

One solves a problem by finding new or amended theories, containing explanations which do not have the deficiencies, but do retain the merits, of existing explanations (Figure 3.2). Thus, after a problem presents itself (stage 1), the next stage always involves *conjecture*: proposing new theories, or modifying or reinterpreting old ones, in the hope of solving the problem (stage 2). The conjectures are then *criticized* which, if the criticism is rational, entails examining and comparing them to see which offers the best explanations, according to the criteria inherent in the problem (stage 3). When a conjectured theory fails to survive criticism – that is, when it appears to offer worse explanations than other theories do – it is abandoned. If we find ourselves abandoning one of our originally held theories in favour of one of the newly proposed ones (stage 4), we tentatively deem our problem-solving enterprise to have made progress. I say 'tentatively', because subsequent problem-solving will probably involve altering or replacing even these new, apparently satisfactory theories, and sometimes even resurrecting some of the apparently unsatisfactory ones. Thus the solution, however good, is not the end of the story: it is a starting-point for the next problem-solving process (stage 5). This illustrates another of the misconceptions behind inductivism. In science the object of the exercise is not to find a theory that will, or is likely to, be deemed true for ever; it is to find the best theory available now, and if possible to improve on all available theories. A scientific argument is intended to persuade us that a given explanation is the best one available. It does not and could not say anything about how that explanation will fare when, in the future, it is subjected to new types of criticism and compared with explanations that have yet to be invented. A good explanation may make good predic-

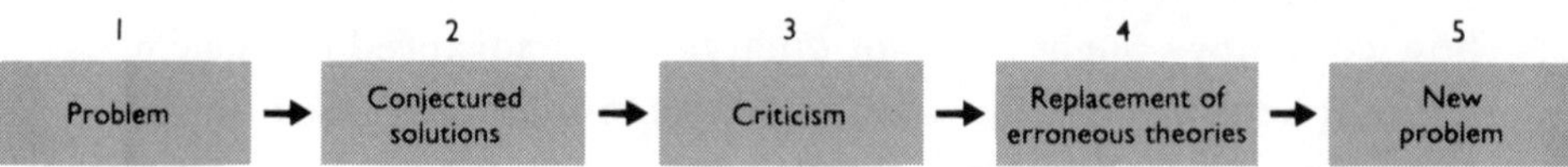

FIGURE 3.2 *The problem-solving process.*

tions about the future, but the one thing that no explanation can even begin to predict is the content or quality of its own future rivals.

What I have described so far applies to all problem-solving, whatever the subject-matter or techniques of rational criticism that are involved. *Scientific* problem-solving always includes a particular method of rational criticism, namely *experimental testing*. Where two or more rival theories make conflicting predictions about the outcome of an experiment, the experiment is performed and the theory or theories that made false predictions are abandoned. The very construction of scientific conjectures is focused on finding explanations that have experimentally testable predictions. Ideally we are always seeking *crucial experimental tests* – experiments whose outcomes, whatever they are, will falsify one or more of the contending theories. This process is illustrated in Figure 3.3. Whether or not observations were involved in the instigating problem (stage 1), and whether or not (in stage 2) the contending theories were specifically designed to be tested experimentally, it is in this critical phase of scientific discovery (stage 3) that experimental tests play their decisive and characteristic role. That role is to render some of the contending theories unsatisfactory by revealing that their explanations lead to false predictions. Here I must mention an asymmetry which is important in the philosophy and methodology of science: the asymmetry between experimental refutation and experimental confirmation. Whereas an incorrect prediction automatically renders the underlying explanation unsatisfactory, a correct prediction says nothing at all about the underlying explanation. Shoddy explanations that yield correct predictions are two a penny, as UFO enthusiasts, conspiracy-theorists and pseudo-scientists of every variety should (but never do) bear in mind.

If a theory about observable events is untestable – that is, if no

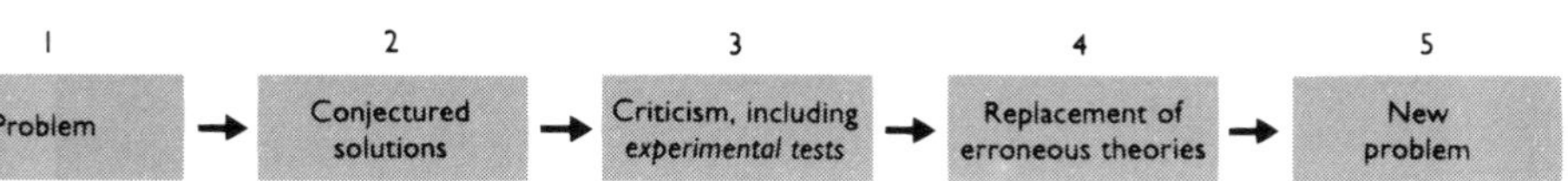

FIGURE 3.3 *The course of scientific discovery.*

possible observation would rule it out – then it cannot by itself explain why those events happen in the way they are observed to and not in some other way. For example, the 'angel' theory of planetary motion is untestable because no matter how planets moved, that motion could be attributed to angels; therefore the angel theory cannot explain the particular motions that we see, unless it is supplemented by an independent theory of how angels move. That is why there is a methodological rule in science which says that once an experimentally testable theory has passed the appropriate tests, any *less* testable rival theories about the same phenomena are summarily rejected, for their explanations are bound to be inferior. This rule is often cited as distinguishing science from other types of knowledge-creation. But if we take the view that science is about explanations, we see that this rule is really a special case of something that applies naturally to all problem-solving: *theories that are capable of giving more detailed explanations are automatically preferred.* They are preferred for two reasons. One is that a theory that 'sticks its neck out' by being more specific about more phenomena opens up itself and its rivals to more forms of criticism, and therefore has more chance of taking the problem-solving process forward. The second is simply that, if such a theory survives the criticism, it leaves less unexplained – which is the object of the exercise.

I have already remarked that even in science most criticism does not consist of experimental testing. That is because most scientific criticism is directed not at a theory's predictions but directly at the underlying explanations. Testing the predictions is just an indirect way (albeit an exceptionally powerful one, when available) of testing the explanations. In Chapter 1, I gave the example of the 'grass cure' – the theory that eating a kilogram of grass is a cure for the common cold. That theory and an infinity of others of the same ilk are readily testable. But we can criticize and reject them without bothering to do any experiments, purely on the grounds that they explain no more than the prevailing theories which they contradict, yet make new, unexplained assertions.

The stages of a scientific discovery shown in Figure 3.3 are seldom completed in sequence at the first attempt. There is usually repeated backtracking before each stage is completed – or rather, *solved*, for each stage may present a problem whose solution itself requires all five stages of a subsidiary problem-solving process. This applies even to stage 1, for the initiating problem itself is not immutable. If we cannot think of good candidate solutions we may return to stage 1 and try to reformulate the problem, or even choose a different problem. Indeed, apparent insolubility is only one of many reasons why we often find it desirable to modify problems we are solving. Some variants of a problem are inevitably more interesting, or more relevant to other problems; some are better formulated; some seem to be potentially more fruitful, or more urgent – or whatever. In many cases the issue of what precisely the problem is, and what the attributes of a 'good' explanation would be, receive as much criticism and conjecture as do trial solutions.

Similarly, if our criticisms at stage 3 fail to distinguish between rival theories, we try to invent new methods of criticism. If that does not seem to work we may backtrack to stage 2 and try to sharpen our proposed solutions (and existing theories) so as to get more explanations and predictions out of them and make it easier to find fault with them. Or we may again backtrack to stage 1 and try to find better criteria for the explanations to meet. And so on.

Not only is there constant backtracking, but the many sub-problems all remain simultaneously active and are addressed opportunistically. It is only when the discovery is complete that a fairly sequential argument, in a pattern something like Figure 3.3, can be presented. It can begin with the latest and best version of the problem; then it can show how some of the rejected theories fail criticism; then it can set out the winning theory, and say why it survives criticism; then it can explain how one copes without the superseded theory; and finally it can point out some of the new problems that this discovery creates or allows for.

While a problem is still in the process of being solved we are dealing with a large, heterogeneous set of ideas, theories, and criteria, with many variants of each, all competing for survival. There is a continual turnover of theories as they are altered or replaced by new ones. So all the theories are being subjected to *variation* and *selection*, according to criteria which are themselves subject to variation and selection. The whole process resembles biological evolution. A problem is like an ecological niche, and a theory is like a gene or a species which is being tested for viability in that niche. Variants of theories, like genetic mutations, are continually being created, and less successful variants become extinct when more successful variants take over. 'Success' is the ability to survive repeatedly under the selective pressures – criticism – brought to bear in that niche, and the criteria for that criticism depend partly on the physical characteristics of the niche and partly on the attributes of other genes and species (i.e. other ideas) that are already present there. The new world-view that may be implicit in a theory that solves a problem, and the distinctive features of a new species that takes over a niche, are *emergent* properties of the problem or niche. In other words, obtaining solutions is inherently complex. There is no simple way of discovering the true nature of planets, given (say) a critique of the celestial-sphere theory and some additional observations, just as there is no simple way of designing the DNA of a koala bear, given the properties of eucalyptus trees. Evolution, or trial and error – especially the focused, purposeful form of trial and error called scientific discovery – are the only ways.

For this reason, Popper has called his theory that knowledge can grow only by conjecture and refutation, in the manner of Figure 3.3, an *evolutionary epistemology*. This is an important unifying insight, and we shall see that there are other connections between these two strands. But I do not want to overstate the similarities between scientific discovery and biological evolution, for there are important differences too. One difference is that in biology variations (mutations) are random, blind and purposeless, while in human problem-solving the creation of new conjectures is itself a complex, knowledge-laden process driven by the intentions of the people concerned. Perhaps an even more important difference is that there is no biological equivalent of *argument*. All conjectures have to be tested experimentally, which is one reason why biological evolution is slower and less efficient by an astronomically large factor. Nevertheless, the link between the two sorts of process is far more than mere analogy: they are two of my four intimately related 'main strands' of explanation of the fabric of reality.

Both in science and in biological evolution, evolutionary success depends on the creation and survival of *objective knowledge*, which in biology is called *adaptation*. That is, the ability of a theory or gene to survive in a niche is not a haphazard function of its structure but depends on whether enough true and useful information about the niche is implicitly or explicitly encoded there. I shall say more about this in Chapter 8.

We can now begin to see what justifies the inferences that we draw from observations. We never draw inferences from observations alone, but observations can become significant in the course of an argument when they reveal deficiencies in some of the contending explanations. We choose a scientific theory because arguments, only a few of which depend on observations, have satisfied us (for the moment) that the explanations offered by all known rival theories are less true, less broad or less deep.

Take a moment to compare Figures 3.1 and 3.3. Look how different these two conceptions of the scientific process are. Inductivism is observation- and prediction-based, whereas in reality science is problem- and explanation-based. Inductivism supposes that theories are somehow extracted or distilled from observations, or are justified by them, whereas in fact theories begin as unjustified conjectures in someone's mind, which typically *precede* the observations that rule out rival theories. Inductivism seeks to justify predictions as likely to hold in the future. Problem-solving justifies an explanation as being better than other explanations available in the present. Inductivism is a dangerous and recurring source of many sorts of error, because it is superficially so plausible. But it is not true.

When we succeed in solving a problem, scientific or otherwise, we end up with a set of theories which, though they are not problem-free, we find preferable to the theories we started with.

What new attributes the new theories will have therefore depends on what we saw as the deficiencies in our original theories – that is, on what the problem was. Science is characterized by its problems as well as by its method. Astrologers who solve the problem of how to cast more intriguing horoscopes without risking being proved wrong are unlikely to have created much that deserves to be called scientific knowledge, even if they have used genuine scientific methods (such as market research) and are themselves quite satisfied with the solution. The problem in genuine science is always to understand some aspect of the fabric of reality, by finding explanations that are as broad and deep, and as true and specific, as possible.

When we think that we have solved a problem, we naturally adopt our new set of theories in preference to the old set. That is why science, regarded as explanation-seeking and problem-solving, raises no 'problem of induction'. There is no mystery about why we should feel compelled tentatively to accept an explanation when it is the best explanation we can think of.

TERMINOLOGY

solipsism The theory that only one mind exists and that what appears to be external reality is only a dream taking place in that mind.

problem of induction Since scientific theories cannot be logically justified by observation, what does justify them?

induction A fictitious process by which general theories were supposed to be obtained from, or justified by, accumulated observations.

problem A problem exists when it seems that some of our theories, especially the explanations they contain, seem inadequate and worth trying to improve.

criticism Rational criticism compares rival theories with the aim of finding which of them offers the best explanations according to the criteria inherent in the problem.

science The *purpose* of science is to understand reality through explanations. The characteristic (though not the only) *method of criticism* used in science is experimental testing.

experimental test An experiment whose outcome may falsify one or more of a set of rival theories.

SUMMARY

In fundamental areas of science, observations of ever smaller, more subtle effects are driving us to ever more momentous conclusions about the nature of reality. Yet these conclusions cannot be deduced by pure logic from the observations. So what makes them compelling? This is the 'problem of induction'. According to inductivism, scientific theories are discovered by extrapolating the results of observations, and justified when corroborating observations are obtained. In fact, inductive reasoning is invalid, and it is impossible to extrapolate observations unless one already has an explanatory framework for them. But the refutation of inductivism, and also the real solution of the problem of induction, depends on recognizing that science is a process not of deriving predictions from observations, but of finding explanations. We seek explanations when we encounter a problem with existing ones. We then embark on a problem-solving process. New explanatory theories begin as unjustified conjectures, which are criticized and compared according to the criteria inherent in the problem. Those that fail to survive this criticism are abandoned. The survivors become the new prevailing theories, some of which are themselves problematic and so lead us to seek even better explanations. The whole process resembles biological evolution.

Thus we acquire ever more knowledge of reality by solving problems and finding better explanations. But when all is said and done, problems and explanations are located within the human mind, which owes its reasoning power to a fallible brain, and its supply of information to fallible senses. What, then, entitles a human mind to draw conclusions about objective, external reality from its own purely subjective experience and reason?

4

Criteria for Reality

The great physicist Galileo Galilei, who was arguably also the first physicist in the modern sense, made many discoveries not only in physics itself but also in the methodology of science. He revived the ancient idea of expressing general theories about nature in mathematical form, and improved upon it by developing the method of systematic experimental testing, which characterizes science as we know it. He aptly called such tests *cimenti*, or 'ordeals'. He was one of the first to use telescopes to study celestial objects, and he collected and analysed evidence for the heliocentric theory, the theory that the Earth moves in orbit around the Sun and spins about its own axis. He is best known for his advocacy of that theory, and for the bitter conflict with the Church into which that advocacy brought him. In 1633 the Inquisition tried him for heresy, and forced him under the threat of torture to kneel and read aloud a long, abject recantation saying that he 'abjured, cursed and detested' the heliocentric theory. (Legend has it, probably incorrectly, that as he rose to his feet he muttered the words *'eppur si muove . . .'*, meaning 'and yet, it does move . . .'.) Despite his recantation, he was convicted and sentenced to house arrest, under which he remained for the rest of his life. Although this punishment was comparatively lenient, it achieved its purpose handsomely. As Jacob Bronowski put it:

> The result was silence among Catholic scientists everywhere from then on . . . The effect of the trial and of the imprisonment was to put a total stop to the scientific tradition in the Mediterranean. (*The Ascent of Man*, p. 218)

How could a dispute about the layout of the solar system have such far-reaching consequences, and why did the participants pursue it so passionately? Because the real dispute was not about whether the solar system had one layout rather than another: it was about Galileo's brilliant advocacy of a new and dangerous way of thinking about reality. Not about the existence of reality, for both Galileo and the Church believed in *realism*, the common-sense view that an external physical universe really does exist and does affect our senses, including senses enhanced by instruments such as telescopes. Where Galileo differed was in his conception of the relationship between physical reality on the one hand, and human ideas, observations and reason on the other. He believed that the universe could be understood in terms of universal, mathematically formulated laws, and that reliable knowledge of these laws was accessible to human beings if they applied his method of mathematical formulation and systematic experimental testing. As he put it, 'the Book of Nature is written in mathematical symbols'. This was in conscious comparison with that other Book on which it was more conventional to rely.

Galileo understood that if his method was indeed reliable, then wherever it was applicable its conclusions had to be preferable to those obtained by any other method. Therefore he insisted that scientific reasoning took precedence not only over intuition and common sense, but also over religious doctrine and revelation. It was specifically that idea, and not the heliocentric theory as such, that the authorities considered dangerous. (And they were right, for if any idea can be said to have initiated the scientific revolution and the Enlightenment, and to have provided the secular foundation of modern civilization, it is that one.) It was forbidden to 'hold or defend' the heliocentric theory *as an explanation* of the appearance of the night sky. But using the heliocentric theory, writing about it, holding it 'as a mathematical supposition' or defending it as a method of making predictions were all permitted. That was why Galileo's book *Dialogue of the Two Chief World Systems*, which compared the heliocentric theory with the official geocentric theory, had been cleared for printing by the Church censors. The Pope had even acquiesced in advance to Galileo's writing such a book (though at the trial a misleading document was produced, claiming that Galileo had been forbidden to discuss the issue at all).

It is an interesting historical footnote that in Galileo's time it was not yet indisputable that the heliocentric theory gave better predictions than the geocentric theory. The available observations were not very accurate. *Ad hoc* modifications had been proposed to improve the accuracy of the geocentric theory, and it was hard to quantify the predictive powers of the two rival theories. Furthermore, when it comes to details, there is more than one heliocentric theory. Galileo believed that the planets move in circles, while in fact their orbits are very nearly ellipses. So the data did not fit the *particular* heliocentric theory that Galileo was defending either. (So much, then, for his having been convinced by accumulated observations!) But for all that, the Church took no position in this controversy. The Inquisition did not care where the planets *appeared* to be; what they cared about was reality. They cared where the planets really were, and they wanted to understand the planets through explanations, just as Galileo did. Instrumentalists and positivists would say that since the Church was perfectly willing to accept Galileo's observational predictions, further argument between them was pointless, and that his muttering '*eppur si muove*' was strictly meaningless. But Galileo knew better, and so did the Inquisition. When they denied the reliability of scientific knowledge, it was precisely the explanatory part of that knowledge that they had in mind.

Their world-view was false, but it was not illogical. Admittedly they believed in revelation and traditional authority as sources of reliable knowledge. But they also had an independent reason for criticizing the reliability of knowledge obtained by Galileo's methods. They could simply point out that no amount of observation or argument can ever prove that one explanation of a physical phenomenon is true and another false. As they would put it, God could produce the same observed effects in an infinity of different ways, so it is pure vanity and arrogance to claim to possess a way of knowing, merely through one's own fallible observation and reason, which way He chose.

To some extent they were merely arguing for modesty, for a recognition of human fallibility. And if Galileo was claiming that the heliocentric theory was somehow proven, or nearly so, in some inductive sense, they had a point. If Galileo thought that his methods could confer on any theory an authority comparable to that which the Church claimed for its doctrines, they were right to criticize him as arrogant (or, as they would have put it, blasphemous), though of course by the same standard they were much more arrogant themselves.

So how can we defend Galileo against the Inquisition? What should Galileo's defence have been in the face of this charge of claiming too much when he claimed that scientific theories contain reliable knowledge of reality? The Popperian defence of science as a process of problem-solving and explanation-seeking is not sufficient in itself. For the Church too was primarily interested in explanations and not predictions, and it was quite willing to let Galileo solve problems using any theory he chose. It was just that they did not accept that Galileo's solutions (which they would call mere 'mathematical hypotheses') had any bearing on external reality. Problem-solving, after all, is a process that takes place entirely within human minds. Galileo may have seen the world as a book in which the laws of nature are written in mathematical symbols. But that is strictly a metaphor; there are no explanations in orbit out there with the planets. The fact is that all our problems and solutions are located within ourselves, having been created by ourselves. When we solve problems in science we arrive through argument at theories whose explanations seem best to us. So, without in any way denying that it is right and proper, and useful, for us to solve problems, the Inquisition and modern sceptics might legitimately ask what scientific problem-solving has to do with reality. We may find our 'best explanations' psychologically satisfying. We may find them helpful in making predictions. We certainly find them essential in every area of technological creativity. All this does justify our continuing to seek them and to use them in those ways. But why should we be obliged to take them as fact? The proposition that the Inquisition forced Galileo to endorse was in effect this: that the Earth is in fact at rest, with the Sun and planets in motion around it; but that the paths on which these astronomical bodies travel are laid out in a complex way which, when viewed from the vantage-point of the Earth, is also consistent with the Sun being at rest and the Earth and planets being in motion. Let me call that the 'Inquisition's theory' of the solar system. If the Inquisition's theory were true, we should still expect the heliocentric theory to make accurate predictions of the results of all Earth-based astronomical observations, even though it would be factually false. It would therefore seem that any observations that appear to support the heliocentric theory lend equal support to the Inquisition's theory.

One could extend the Inquisition's theory to account for more detailed observations that support the heliocentric theory, such as observations of the phases of Venus, and of the small additional motions (called 'proper motions') of some stars relative to the celestial sphere. To do this one would have to postulate even more complex manoeuvrings in space, governed by laws of physics very different from those that operate on our supposedly stationary Earth. But they would be different in precisely such a way as to remain observationally consistent with the Earth being in motion and the laws being the same out there as they are here. Many such theories are possible. Indeed, if making the right predictions were our only constraint, we could invent theories which say that anything we please is going on in space. For example, observations alone can never rule out the theory that the Earth is enclosed in a giant planetarium showing us a simulation of a heliocentric solar system; and that outside the planetarium there is anything you like, or nothing at all. Admittedly, to account for present-day observations the planetarium would also have to redirect our radar and laser pulses, capture our space probes, and indeed astronauts, send back fake messages from them and return them with appropriate moonrock samples, altered memories, and so on. It may be an absurd theory, but the point is that it cannot be ruled out by experiment. Nor is it valid to rule out any theory solely on the grounds that it is 'absurd': the Inquisition, together with most of the human race in Galileo's time, thought it the epitome of absurdity to claim that the Earth is moving. After all, we cannot feel it moving, can we? When it does move, as in an earthquake, we feel that unmistakably. It is said that Galileo delayed publicly advocating the heliocentric theory for some years, not for fear of the Inquisition but simply for fear of ridicule.

To us, the Inquisition's theory looks hopelessly contrived. Why should we accept such a complicated and *ad hoc* account of why the sky looks as it does, when the unadorned heliocentric cosmology does the same job with less fuss? We may cite the principle of Occam's razor: 'do not multiply entities beyond necessity' – or, as I prefer to put it, 'do not complicate explanations beyond necessity', because if you do, the unnecessary complications themselves remain unexplained. However, whether an explanation is or is not 'contrived' or 'unnecessarily complicated' depends on all the other ideas and explanations that make up one's world-view. The Inquisition would have argued that the idea of the Earth moving is an unnecessary complication. It contradicts common sense; it contradicts Scripture; and (they would have said) there is a perfectly good explanation that does without it.

But is there? Does the Inquisition's theory really provide alternative explanations without having to introduce the counter-intuitive 'complication' of the heliocentric system? Let us take a closer look at how the Inquisition's theory explains things. It explains the apparent stationarity of the Earth by saying that it *is* stationary. So far, so good. On the face of it that explanation is better than Galileo's, for he had to work very hard, and contradict some common-sense notions of force and inertia, to explain why we do not feel the Earth move. But how does the Inquisition's theory cope with the more difficult task of explaining planetary motions?

The heliocentric theory explains them by saying that the planets are seen to move in complicated loops across the sky because they are really moving in simple circles (or ellipses) in space, but the Earth is moving as well. The Inquisition's explanation is that the planets are seen to move in complicated loops because they really are moving in complicated loops in space; *but* (and here, according to the Inquisition's theory, comes the essence of the explanation) this complicated motion is governed by a simple underlying principle: namely, that the planets move in such a way that, when viewed from the Earth, they appear just as they would if they and the Earth were in simple orbits round the Sun.

To understand planetary motions in terms of the Inquisition's theory, it is essential that one should understand this principle, for the constraints it imposes are the basis of every detailed explanation that one can make under the theory. For example, if one were asked why a planetary conjunction occurred on such-and-such a date, or why a planet backtracked across the sky in a loop of a particular shape, the answer would always be 'because that is how it would look if the heliocentric theory were true'. So here is a cosmology – the Inquisition's cosmology – that can be understood only in terms of a different cosmology, the heliocentric cosmology that it contradicts but faithfully mimics.

If the Inquisition had seriously tried to understand the world in terms of the theory they tried to force on Galileo, they would also have understood its fatal weakness, namely that it fails to solve the problem it purports to solve. It does *not* explain planetary motions 'without having to introduce the complication of the heliocentric system'. On the contrary, it unavoidably incorporates that system as part of its own principle for explaining planetary motions. One cannot understand the world through the Inquisition's theory unless one understands the heliocentric theory first.

Therefore we are right to regard the Inquisition's theory as a convoluted elaboration of the heliocentric theory, rather than vice versa. We have arrived at this conclusion not by judging the Inquisition's theory against modern cosmology, which would have been a circular argument, but by insisting on taking the Inquisition's theory seriously, in its own terms, as an explanation of the world. I have mentioned the grass-cure theory, which can be ruled out without experimental testing because it contains no explanation. Here we have a theory which can also be ruled out without experimental testing, because it contains a bad explanation – an explanation which, in its own terms, is worse than its rival.

As I have said, the Inquisition were realists. Yet their theory has this in common with solipsism: both of them draw an arbitrary boundary beyond which, they claim, human reason has no access – or at least, beyond which problem-solving is no path to understanding. For solipsists, the boundary tightly encloses their own brains, or perhaps just their abstract minds or incorporeal souls. For the Inquisition, it enclosed the entire Earth. Some present-day Creationists believe in a similar boundary, not in space but in time, for they believe that the universe was created only six thousand years ago, complete with misleading evidence of earlier events. *Behaviourism* is the doctrine that it is not meaningful to explain human behaviour in terms of inner mental processes. To behaviourists, the only legitimate psychology is the study of people's observable responses to external stimuli. Thus they draw exactly the same boundary as solipsists, separating the human mind from external reality; but while solipsists deny that it is meaningful to reason about anything outside that boundary, behaviourists deny that it is meaningful to reason about anything inside.

There is a large class of related theories here, but we can usefully regard them all as variants of solipsism. They differ in where they draw the boundary of reality (or the boundary of that part of reality which is comprehensible through problem-solving), and they differ in whether, and how, they seek knowledge outside that boundary. But they all consider scientific rationality and other problem-solving to be inapplicable outside the boundary – a mere game. They might concede that it can be a satisfying and useful game, but it is nevertheless only a game from which no valid conclusion can be drawn about the reality outside.

They are also alike in their basic objection to problem-solving as a means of creating knowledge, which is that it does not deduce its conclusions from any ultimate source of justification. Within the respective boundaries that they choose, the adherents of all these theories do rely on the methodology of problem-solving, confident that seeking the best available explanation is also the way of finding the truest available theory. But for the truth of what lies outside those boundaries, they look elsewhere, and what they all seek is a source of ultimate justification. For religious people, divine revelation can play that role. Solipsists trust only the direct experience of their own thoughts, as expressed in René Descartes's classic argument *cogito ergo sum* ('I think, therefore I exist').

Despite Descartes's desire to base his philosophy on this supposedly firm foundation, he actually allowed himself many other assumptions, and he was certainly no solipsist. Indeed, there can have been very few, if any, genuine solipsists in history. Solipsism is usually defended only as a means of attacking scientific reasoning, or as a stepping-stone to one of its many variants. By the same token, a good way of defending science against a variety of criticisms, and of understanding the true relationship between reason and reality, is to consider the argument against solipsism.

There is a standard philosophical joke about a professor who gives a lecture in defence of solipsism. So persuasive is the lecture that as soon as it ends, several enthusiastic students hurry forward to shake the professor's hand. 'Wonderful. I agreed with every word,' says one student earnestly. 'So did I,' says another. 'I am very gratified to hear it,' says the professor. 'One so seldom has the opportunity to meet fellow solipsists.'

Implicit in this joke there is a genuine argument against solipsism. One could put it like this. What, exactly, was the theory that the students in the story were agreeing with? Was it the professor's theory, that they themselves do not exist because only the professor exists? To believe that, they would first have had to find some way round Descartes's *cogito ergo sum* argument. And if they managed that, they would not be solipsists, for the central thesis of solipsism is that the solipsist exists. Or has each student been persuaded of a theory *contradicting* the professor's, the theory that that particular student exists, but the professor and the other students do not? That would indeed make them all solipsists, but none of the students would be agreeing with the theory that the professor was defending. Therefore neither of these two possibilities amounts to the students' having been persuaded by the professor's defence of solipsism. If they adopt the professor's opinion, they will not be solipsists, and if they become solipsists, they will have become convinced that the professor is mistaken.

This argument is trying to show that solipsism is literally indefensible, because by accepting such a defence one is implicitly contradicting it. But our solipsistic professor could try to evade that argument by saying something like this: 'I can and do consistently defend solipsism. Not against other people, for there are no other people, but against opposing arguments. These arguments come to my attention through dream-people, who behave as if they were thinking beings whose ideas often oppose mine. My lecture and the arguments it contains were not intended to persuade these dream-people, but to persuade myself – to help me to clarify my ideas.'

However, if there are sources of ideas that behave *as if* they were independent of oneself, then they necessarily *are* independent of oneself. For if I define 'myself' as the conscious entity that has the thoughts and feelings I am aware of having, then the 'dream-people' I seem to interact with are by definition something other than that narrowly defined self, and so I must concede that something other than myself exists. My only other option, if I were a committed solipsist, would be to regard the dream-people as creations of my unconscious mind, and therefore as part of 'myself' in a looser sense. But then I should be forced to concede that 'myself' had a very rich structure, most of which is independent of my conscious self. Within that structure are entities – dream-people – who, despite being mere constituents of the mind of a supposed solipsist, behave exactly as if they were committed *anti*-solipsists. So I could not call myself wholly a solipsist, for only my narrowly defined self would take that view. Many, apparently most, of the opinions held within my mind as a whole would oppose solipsism. I could study the 'outer' region of myself and find that it seems to obey certain laws, the same laws as the dream-textbooks say apply to what they call the physical universe. I would find that there is far more of the outer region than the inner region. Aside from containing more ideas, it is also more complex, more varied, and has more measurable variables, by a literally astronomical factor, than the inner region.

Moreover, this outer region is amenable to scientific study, using the methods of Galileo. Because I have now been forced to define that region as part of myself, solipsism no longer has any argument against the validity of such study, which is now defined as no more than a form of introspection. Solipsism allows, indeed assumes, that knowledge of oneself can be obtained through introspection. It cannot declare the entities and processes being studied to be unreal, since the reality of the self is its basic postulate.

Thus we see that if we take solipsism seriously – if we assume that it is true and that all valid explanations must scrupulously conform to it – it self-destructs. How exactly does solipsism, taken seriously, differ from its common-sense rival, realism? The difference is based on no more than a renaming scheme. Solipsism insists on referring to objectively different things (such as external reality and my unconscious mind, or introspection and scientific observation) by the same names. But then it has to reintroduce the distinction through explanations in terms of something like the 'outer part of myself'. But no such extra explanations would be necessary without its insistence on an inexplicable renaming scheme. Solipsism must also postulate the existence of an additional class of processes – invisible, inexplicable processes which give the mind the illusion of living in an external reality. The solipsist, who believes that nothing exists other than the contents of one mind, must also believe that that mind is a phenomenon of greater multiplicity than is normally supposed. It contains other-people-like thoughts, planet-like thoughts and laws-of-physics-like thoughts. These thoughts are real. They develop in a complex way (or pretend to), and they have enough autonomy to surprise, disappoint, enlighten or thwart that other class of thoughts which call themselves 'I'. Thus the solipsist's explanation of the world is in terms of interacting thoughts rather than interacting objects. But those thoughts are real, and interact according to the same rules that the realist says govern the interaction of objects. Thus solipsism, far from being a world-view stripped to its essentials, is actually just realism disguised and weighed down by additional unnecessary assumptions – worthless baggage, introduced only to be explained away.

By this argument we can dispense with solipsism and all the related theories. They are all indefensible. Incidentally, we have already rejected one world-view on these grounds, namely positivism (the theory that all statements other than those describing or predicting observations are meaningless). As I remarked in Chapter 1, positivism asserts its own meaninglessness, and therefore cannot be consistently defended.

So we can continue, reassured, with common-sense realism and the pursuit of explanations by scientific methods. But in the light of this conclusion, what can we say about the arguments that made solipsism and its relatives superficially plausible, namely that they could neither be proved false nor ruled out by experiment? What is the status of those arguments now? If we have neither proved solipsism false nor ruled it out by experiment, what *have* we done?

There is an assumption built into this question. It is that theories can be classified in a hierarchy, 'mathematical' → 'scientific' → 'philosophical', of decreasing intrinsic reliability. Many people take the existence of this hierarchy for granted, despite the fact that these judgements of comparative reliability depend entirely on philosophical arguments, arguments that classify themselves as quite unreliable! In fact, the idea of this hierarchy is a cousin of the reductionist mistake I discussed in Chapter 1 (the theory that microscopic laws and phenomena are more fundamental than emergent ones). The same assumption occurs in inductivism, which supposes that we can be absolutely certain of the conclusions of mathematical arguments because they are deductive, reasonably sure of scientific arguments because they are 'inductive', and forever undecided about philosophical arguments, which it sees as little more than matters of taste.

But none of that is true. Explanations are not justified by the means by which they were derived; they are justified by their superior ability, relative to rival explanations, to solve the problems they address. That is why the argument that a theory is *indefensible* can be so compelling. A prediction, or any assertion, that cannot be defended might still be true, but an explanation that cannot be defended is not an explanation. The rejection of 'mere' explanations on the grounds that they are not justified by any *ultimate* explanation inevitably propels one into futile searches for an ultimate source of justification. There is no such source.

Nor is there that hierarchy of reliability from mathematical to scientific to philosophical arguments. Some philosophical arguments, including the argument against solipsism, are far more compelling than any scientific argument. Indeed, every scientific argument assumes the falsity not only of solipsism, but also of other philosophical theories including any number of variants of solipsism that might contradict specific parts of the scientific argument. I shall also show (in Chapter 10) that even purely mathematical arguments derive their reliability from the physical and philosophical theories that underpin them, and therefore that they cannot, after all, yield absolute certainty.

Having embraced realism, we are continually faced with decisions as to whether entities referred to in competing explanations are real or not. Deciding that they are not real – as we did in the case of the 'angel' theory of planetary motion – is equivalent to rejecting the corresponding explanation. Thus, in searching for and judging explanations, we need more than just a refutation of solipsism. We need to develop reasons for accepting or rejecting the existence of entities that may appear in contending theories; in other words, we need a criterion for reality. We should not, of course, expect to find a final or an infallible criterion. Our judgements of what is or is not real always depend on the various explanations that are available to us, and sometimes change as our explanations improve. In the nineteenth century, few things would have been regarded more confidently as real than the force of gravity. Not only did it figure in Newton's then-unrivalled system of laws, but everyone could feel it, all the time, even with their eyes shut – or so they thought. Today we understand gravity through Einstein's theory rather than Newton's, and we know that no such force exists. We do *not* feel it! What we feel is the resistance that prevents us from penetrating the solid ground beneath our feet. Nothing is pulling us downwards. The only reason why we fall downwards when unsupported is that the fabric of space and time in which we exist is curved.

Not only do explanations change, but our criteria and ideas about what should count as an explanation are gradually changing (improving) too. So the list of acceptable modes of explanation will always be open-ended, and consequently the list of acceptable criteria for reality must be open-ended too. But what is it about an explanation – given that, for whatever reasons, we find it satisfactory – that should make us classify some things as real and others as illusory or imaginary?

James Boswell relates in his *Life of Johnson* how he and Dr Johnson were discussing Bishop Berkeley's solipsistic theory of the non-existence of the material world. Boswell remarked that although no one believed the theory, no one could refute it either. Dr Johnson kicked a large rock and said, as his foot rebounded, 'I refute it *thus*.' Dr Johnson's point was that Berkeley's denial of the rock's existence is incompatible with finding an explanation of the rebound that he himself felt. Solipsism cannot accommodate any explanation of why that experiment – or any experiment – should have one outcome rather than another. To explain the effect that the rock had on him, Dr Johnson was forced to take a position on the nature of rocks. Were they part of an autonomous external reality, or were they figments of his imagination? In the latter case he would have to conclude that 'his imagination' was itself a vast, complex, autonomous universe. The same dilemma confronted the solipsistic professor who, if pressed for explanations, would be forced to take a position on the nature of the audience. And the Inquisition would have had to take a position on the source of the underlying regularity in the motion of planets, a regularity that is explicable only by reference to the heliocentric theory. For all these people, taking their own position seriously as an explanation of the world would lead them directly to realism and Galilean rationality.

But Dr Johnson's idea is more than a refutation of solipsism. It also illustrates the criterion for reality that is used in science, namely, *if something can kick back, it exists*. 'Kicking back' here does not necessarily mean that the alleged object is responding to being kicked – to being physically affected as Dr Johnson's rock was. It is enough that when we 'kick' something, the object affects us in ways that require independent explanation. For example, Galileo had no means of affecting planets, but he could affect the light that came from them. His equivalent of kicking the rock was refracting that light through the lenses of his telescopes and eyes. That light responded by 'kicking' his retina back. The way it kicked back allowed him to conclude not only that the light was real, but that the heliocentric planetary motions required to explain the patterns in which the light arrived were also real.

By the way, Dr Johnson did not directly kick the rock either. A person is a mind, not a body. The Dr Johnson who performed the experiment was a mind, and that mind directly 'kicked' only some nerves, which transmitted signals to muscles, which propelled his foot towards the rock. Shortly afterwards, Dr Johnson perceived being 'kicked back' by the rock, but again only indirectly, after the impact had set up a pressure pattern in his shoe, and then in his skin, and had then led to electrical impulses in his nerves, and so forth. Dr Johnson's mind, like Galileo's and everyone else's, 'kicked' nerves and 'was kicked back' by nerves, and inferred the existence and properties of reality from those interactions alone. *What* Dr Johnson was entitled to infer about reality depends on how he could best explain what had happened. For example, if the sensation had seemed to depend only on the extension of his leg, and not on external factors, then he would probably have concluded that it was a property of his leg, or of his mind alone. He might have been suffering from a disease which gave him a rebounding sensation whenever he extended his leg in a certain way. But in fact the rebounding depended on what the rock did, such as being in a certain place, which was in turn related to other effects that the rock had, such as being seen, or affecting other people who kicked it. Dr Johnson perceived these effects to be autonomous (independent of himself) and quite complicated. Therefore the realist explanation of why the rock produces the rebounding sensation involves a complicated story about something autonomous. But so does the solipsist explanation. In fact, any explanation that accounts for the foot-rebounding phenomenon is necessarily a 'complicated story about something autonomous'. It must in effect be the story of the rock. The solipsist would call it a dream-rock, but apart from that claim the solipsist's story and the realist's could share the same script.

My discussion of shadows and parallel universes in Chapter 2 revolved around questions of what does or does not exist, and implicitly around what should or should not count as evidence of existence. I used Dr Johnson's criterion. Consider again the point X on the screen in Figure 2.7 (p. 41), which is illuminated when only two slits are open but goes dark when two further slits are opened. I said that it is an 'inescapable' conclusion that *something* must be coming through the second pair of slits to prevent the light from the first pair from reaching X. It is not *logically* inescapable, for if we were not looking for explanations we could just say that the photons we see behave *as if* something passing through other slits had deflected them, but that in fact there is nothing there. Similarly, Dr Johnson could have said that his foot rebounded *as if* a rock had been there, but that in fact there was nothing there. The Inquisition did say that the planets were seen to move as if they and the Earth were in orbit round the Sun, but that in fact they moved round the fixed Earth. But if the object of the exercise is to explain the motion of planets, or the motion of photons, we must do as Dr Johnson did. We must adopt a methodological rule that if something behaves as if it existed, by kicking back, then one regards that as evidence that it does exist. Shadow photons kick back by interfering with the photons that we see, and therefore shadow photons exist.

Can we likewise conclude from Dr Johnson's criterion that 'planets move as if they were being pushed by angels; therefore angels exist'? No, but only because we have a better explanation. The angel theory of planetary motion is not *wholly* without merit. It does explain why planets move independently of the celestial sphere, and that does indeed make it superior to solipsism. But it does not explain why the angels should push the planets along one set of orbits rather than another, or, in particular, why they should push them as if their motion were determined by a curvature of space and time, as specified in every detail by the universal laws of the general theory of relativity. That is why the angel theory cannot compete as an explanation with the theories of modern physics.

Similarly, to postulate that angels come through the other slits and deflect our photons would be better than nothing. But we can do better than that. We know exactly how those angels would have to behave: very much like photons. So we have a choice between an explanation in terms of invisible angels pretending to be photons, and one in terms of invisible photons. In the absence of an independent explanation for why angels should pretend to be photons, that latter explanation is superior.

We do not feel the presence of our counterparts in other universes. Nor did the Inquisition feel the Earth moving beneath their feet. And yet, it moves! Now, consider what it would feel like if we did exist in multiple copies, interacting only through the imperceptibly slight effects of quantum interference. This is the equivalent of what Galileo did when he analysed how the Earth would feel to us if it were moving in accordance with the heliocentric theory. He discovered that the motion would be imperceptible. Yet perhaps 'imperceptible' is not quite the right word here. Neither the motion of the Earth nor the presence of parallel universes is directly perceptible, but then neither is anything else (except perhaps, if Descartes's argument holds, your own bare existence). But both things *are* perceptible in the sense that they perceptibly 'kick back' at us if we examine them through scientific instruments. We can see a Foucault pendulum swing in a plane that gradually seems to turn, revealing the rotation of the Earth beneath it. And we can detect photons that have been deflected by interference from their other-universe counterparts. It is only an accident of evolution, as it were, that the senses we are born with are not adapted to feel such things 'directly'.

It is not how hard something kicks back that makes the theory of its existence compelling. What matters is its role in the explanations that such a theory provides. I have given examples from physics where very tiny 'kicks' lead us to momentous conclusions about reality because we have no other explanation. The converse can also happen: if there is no clear-cut winner among the contending explanations, then even a very powerful 'kick' may not convince us that the supposed source has independent reality. For example, you may one day see terrifying monsters attacking you – and then wake up. If the explanation that they originated within your own mind seems adequate, it would be irrational for you to conclude that there really are such monsters out there. If you feel a sudden pain in your shoulder as you walk down a busy street, and look around, and see nothing to explain it, you may wonder whether the pain was caused by an unconscious part of your own mind, or by your body, or by something outside. You may consider it *possible* that a hidden prankster has shot you with an air-gun, yet come to no conclusion as to the reality of such a person. But if you then saw an air-gun pellet rolling away on the pavement, you might conclude that no explanation solved the problem as well as the air-gun explanation, in which case you would adopt it. In other words, you would tentatively infer the existence of a person you had not seen, and might never see, just because of that person's role in the best explanation available to you. Clearly the theory of such a person's existence is not a logical consequence of the observed evidence (which, incidentally, would consist of a single observation). Nor does that theory have the form of an 'inductive generalization', for example that you will observe the same thing again if you perform the same experiment. Nor is the theory experimentally testable: experiment could never prove the absence of a hidden prankster. Despite all that, the argument in favour of the theory could be overwhelmingly convincing, if it were the best explanation.

Whenever I have used Dr Johnson's criterion to argue for the reality of something, one attribute in particular has always been relevant, namely *complexity*. We prefer simpler explanations to more complex ones. And we prefer explanations that are capable of accounting for detail and complexity to explanations that can account only for simple aspects of phenomena. Dr Johnson's criterion tells us to regard as real those complex entities which, if we did *not* regard them as real, would complicate our explanations. For instance, we must regard the planets as real, because if we did not we should be forced into complicated explanations of a cosmic planetarium, or of altered laws of physics, or of angels, or of whatever else would, under that assumption, be giving us the illusion that there are planets out there in space.

Thus the observed complexity in the structure or behaviour of an entity is part of the evidence that that entity is real. But it is not sufficient evidence. We do not, for example, deem our reflections in a mirror to be real people. Of course, illusions themselves are real physical processes. But the illusory entities they show us need not be considered real, because they derive their complexity from somewhere else. They are not *autonomously* complex. Why do we accept the 'mirror' theory of reflections, but reject the 'planetarium' theory of the solar system? It is because, given a simple explanation of the action of mirrors, we can understand that nothing of what we see in them genuinely lies behind them. No further explanation is needed because the reflections, though complex, are not autonomous – their complexity is merely borrowed from our side of the mirror. That is not so for planets. The theory that the cosmic planetarium is real, and that nothing lies beyond it, only makes the problem worse. For if we accepted it, then instead of asking only how the solar system works we should first have to ask how the planetarium works, and *then* how the solar system it is displaying works. We could not avoid the latter question, and it is effectively a repetition of what we were trying to answer in the first place. Now we can rephrase Dr Johnson's criterion thus:

If, according to the simplest explanation, an entity is complex and autonomous, then that entity is real.

Computational *complexity theory* is the branch of computer science that is concerned with what resources (such as time, memory capacity or energy) are required to perform given classes of computations. The complexity of a piece of information is defined in terms of the computational resources (such as the length of the program, the number of computational steps or the amount of memory) that a computer would need if it was to reproduce that piece of information. Several different definitions of complexity are in use, each with its own domain of applicability. The exact definitions need not concern us here, but they are all based on the idea that a complex process is one that in effect presents us with the results of a substantial computation. The sense in which the motion of the planets 'presents us with the results of a substantial computation' is well illustrated by a planetarium. Consider a planetarium controlled by a computer which calculates the exact image that the projectors should display to represent the night sky. To do this authentically, the computer has to use the formulae provided by astronomical theories; in fact the computation is identical to the one that it would perform if it were calculating predictions of where an observatory should point its telescopes to see real planets and stars. What we mean by saying that the appearance of the planetarium is 'as complex' as that of the night sky it depicts is that those two computations – one describing the night sky, the other describing the planetarium – are largely identical. So we can re-express Dr Johnson's criterion again, in terms of hypothetical computations:

If a substantial amount of computation would be required to give us the illusion that a certain entity is real, then that entity is real.

If Dr Johnson's leg invariably rebounded when he extended it, then the source of his illusions (God, a virtual-reality machine, or whatever) would need to perform only a simple computation to determine when to give him the rebounding sensation (something like 'IF leg-is-extended THEN rebound . . .'). But to reproduce what Dr Johnson experienced in a realistic experiment it would be necessary to take into account where the rock is, and whether Dr Johnson's foot is going to hit or miss it, and how heavy, how hard and how firmly lodged it is, and whether anyone else has just kicked it out of the way, and so on – a vast computation.

Physicists trying to cling to a single-universe world-view sometimes try to explain quantum interference phenomena as follows: 'No shadow photons exist,' they say, 'and what carries the effect of the distant slits to the photon we see is – nothing. Some sort of action at a distance (as in Newton's law of gravity) simply makes photons change course when a distant slit is opened.' But there is nothing 'simple' about this supposed action at a distance. The appropriate physical law would have to say that a photon is affected by distant objects exactly *as if* something were passing through the distant gaps and bouncing off the distant mirrors so as to intercept that photon at the right time and place. Calculating how a photon reacts to these distant objects would require the same computational effort as working out the history of large numbers of shadow photons. The computation would have to work its way through a story of what each shadow photon does: it bounces off this, is stopped by that, and so on. Therefore, just as with Dr Johnson's rock, and just as with Galileo's planets, a story that is in effect about shadow photons necessarily appears in any explanation of the observed effects. The irreducible complexity of that story makes it philosophically untenable to deny that the objects exist.

The physicist David Bohm constructed a theory with predictions identical to those of quantum theory, in which a sort of wave accompanies every photon, washes over the entire barrier, passes through the slits and interferes with the photon that we see. Bohm's theory is often presented as a single-universe variant of quantum theory. But according to Dr Johnson's criterion, that is a mistake. Working out what Bohm's invisible wave will do requires the same computations as working out what trillions of shadow photons will do. Some parts of the wave describe us, the observers, detecting and reacting to the photons; other parts of the wave describe other versions of us, reacting to photons in different positions. Bohm's modest nomenclature – referring to most of reality as a 'wave' – does not change the fact that in his theory reality consists of large sets of complex entities, each of which can perceive other entities in its own set, but can only indirectly perceive entities in other sets. These sets of entities are, in other words, parallel universes.

I have described Galileo's new conception of our relationship with external reality as a great methodological discovery. It gave us a new, reliable form of reasoning involving observational evidence. That is indeed one aspect of his discovery: scientific reasoning is reliable, not in the sense that it certifies that any particular theory will survive unchanged, even until tomorrow, but in the sense that we are right to rely on it. For we are right to seek solutions to problems rather than sources of ultimate justification. Observational evidence is indeed evidence, not in the sense that any theory can be deduced, induced or in any other way inferred from it, but in the sense that it can constitute a genuine reason for preferring one theory to another.

But there is another side to Galileo's discovery which is much less often appreciated. The reliability of scientific reasoning is not just an attribute of *us*: of *our* knowledge and *our* relationship with reality. It is also a new fact about physical reality itself, a fact which Galileo expressed in the phrase 'the Book of Nature is written in mathematical symbols'. As I have said, it is impossible literally to 'read' any shred of a theory in nature: that is the inductivist mistake. But what is genuinely out there is evidence, or, more precisely, a reality that will respond with evidence if we interact appropriately with it. Given a shred of a theory, or rather, shreds of several rival theories, the evidence is available out there to enable us to distinguish between them. Anyone can search for it, find it and improve upon it if they take the trouble. They do not need authorization, or initiation, or holy texts. They need only be looking in the right way – with fertile problems and promising theories in mind. This open accessibility, not only of evidence but of the whole mechanism of knowledge acquisition, is a key attribute of Galileo's conception of reality.

Galileo may have thought this self-evident, but it is not. It is a substantive assertion about what physical reality is like. Logically, reality need not have had this science-friendly property, but it does – and in abundance. Galileo's universe is saturated with evidence.

Copernicus had assembled evidence for his heliocentric theory in Poland. Tycho Brahe had collected his evidence in Denmark, and Kepler had in Germany. And by pointing his telescope at the skies over Italy, Galileo gained greater access to the same evidence. Every part of the Earth's surface, on every clear night, for billions of years, has been deluged with evidence about the facts and laws of astronomy. For many other sciences evidence has similarly been on display, to be viewed more clearly in modern times by microscopes and other instruments. Where evidence is not already physically present, we can bring it into existence with devices such as lasers and pierced barriers – devices which it is open to anyone, anywhere and at any time, to build. And the evidence will be the same, regardless of who reveals it. The more fundamental a theory is, the more readily available is the evidence that bears upon it (to those who know how to look), not just on Earth but throughout the multiverse.

Thus physical reality is *self-similar* on several levels: among the stupendous complexities of the universe and multiverse, some patterns are nevertheless endlessly repeated. Earth and Jupiter are in many ways dramatically dissimilar planets, but they both move in ellipses, and they are made of the same set of a hundred or so chemical elements (albeit in different proportions), and so are their parallel-universe counterparts. The evidence that so impressed Galileo and his contemporaries also exists on other planets and in distant galaxies. The evidence being considered at this moment by physicists and astronomers would also have been available a billion years ago, and will still be available a billion years hence. The very existence of general, explanatory theories implies that disparate objects and events are physically alike in some ways. The light reaching us from distant galaxies is, after all, only light, but it looks to us like galaxies. Thus reality contains not only evidence, but also the means (such as our minds, and our artefacts) of understanding it. There *are* mathematical symbols in physical reality. The fact that it is we who put them there does not make them any less physical. In those symbols – in our planetariums, books, films and computer memories, and in our brains – there are images of physical reality at large, images not just of the appearance of objects, but of the structure of reality. There are laws and explanations, reductive and emergent. There are descriptions and explanations of the Big Bang and of subnuclear particles and processes; there are mathematical abstractions; fiction; art; morality; shadow photons; parallel universes. To the extent that these symbols, images and theories are true – that is, they resemble in appropriate respects the concrete or abstract things they refer to – their existence gives reality a new sort of self-similarity, the self-similarity we call knowledge.

TERMINOLOGY

heliocentric theory The theory that the Earth moves round the Sun, and spins on its own axis.

geocentric theory The theory that the Earth is at rest and other astronomical bodies move around it.

realism The theory that an external physical universe exists objectively and affects us through our senses.

Occam's razor (My formulation) *Do not complicate explanations beyond necessity*, because if you do, the unnecessary complications themselves will remain unexplained.

Dr Johnson's criterion (My formulation) *If it can kick back, it exists*. A more elaborate version is: *If, according to the simplest explanation, an entity is complex and autonomous, then that entity is real.*

self-similarity Some parts of physical reality (such as symbols, pictures or human thoughts) resemble other parts. The resemblance may be concrete, as when the images in a planetarium resemble the night sky; more importantly, it may be abstract, as when a statement in quantum theory printed in a book correctly explains an aspect of the structure of the multiverse.
(Some readers may be familiar with the geometry of fractals; the notion of self-similarity defined here is much broader than the one used in that field.)

complexity theory The branch of computer science concerned with what resources (such as time, memory capacity or energy) are required to perform given classes of computations.

SUMMARY

Although solipsism and related doctrines are logically self-consistent, they can be comprehensively refuted simply by taking them seriously as explanations. Although they all claim to be simplified world-views, such an analysis shows them to be indefensible over-elaborations of realism. Real entities behave in a complex and autonomous way, which can be taken as the criterion for reality: if something 'kicks back', it exists. Scientific reasoning, which uses observation not as a basis for extrapolation but to distinguish between otherwise equally good explanations, can give us genuine knowledge about reality.

Thus science and other forms of knowledge are made possible by a special self-similarity property of the physical world. Yet it was not physicists who first recognized and studied this property: it was mathematicians and computer theorists, and they called it the universality of computation. *The theory of computation is our third strand.*

5

Virtual Reality

The theory of computation has traditionally been studied almost entirely in the abstract, as a topic in pure mathematics. This is to miss the point of it. Computers are physical objects, and computations are physical processes. What computers can or cannot compute is determined by the laws of physics alone, and not by pure mathematics. One of the most important concepts of the theory of computation is *universality*. A *universal computer* is usually defined as an abstract machine that can mimic the computations of any other abstract machine in a certain well-defined class. However, the significance of universality lies in the fact that universal computers, or at least good approximations to them, can actually be built, and can be used to compute not just each other's behaviour but the behaviour of interesting physical and abstract entities. The fact that this is possible is part of the self-similarity of physical reality that I mentioned in the previous chapter.

The best-known physical manifestation of universality is an area of technology that has been mooted for decades but is only now beginning to take off, namely *virtual reality*. The term refers to any situation in which a person is artificially given the experience of being in a specified environment. For example, a flight simulator – a machine that gives pilots the experience of flying an aircraft without their having to leave the ground – is a type of virtual-reality generator. Such a machine (or more precisely, the computer that controls it) can be programmed with the characteristics of a real or imaginary aircraft. The aircraft's environment, such as the weather and the layout of airports, can also be specified in the program. As the pilot practises flying from one airport to another, the simulator causes the appropriate images to appear at the windows, the appropriate jolts and accelerations to be felt, the corresponding readings to be shown on the instruments, and so on. It can incorporate the effects of, for example, turbulence, mechanical failure and proposed modifications to the aircraft. Thus a flight simulator can give the user a wide range of piloting experiences, including some that no real aircraft could: the simulated aircraft could have performance characteristics that violate the laws of physics: it could, for instance, fly through mountains, faster than light or without fuel.

Since we experience our environment through our senses, any virtual-reality generator must be able to manipulate our senses, overriding their normal functioning so that we can experience the specified environment instead of our actual one. This may sound like something out of Aldous Huxley's *Brave New World*, but of course technologies for the artificial control of human sensory experience have been evolving for thousands of years. All techniques of representational art and long-distance communication may be thought of as 'overriding the normal functioning of the senses'. Even prehistoric cave paintings gave the viewer something of the experience of seeing animals that were not actually there. Today we can do that much more accurately, using movies and sound recordings, though still not accurately enough for the simulated environment to be mistaken for the original.

I shall use the term *image generator* for any device, such as a planetarium, a hi-fi system or a spice rack, which can generate specifiable sensory input for the user: specified pictures, sounds, odours, and so on all count as 'images'. For example, to generate the olfactory image (i.e. the smell) of vanilla, one opens the vanilla bottle from the spice rack. To generate the auditory image (i.e. the sound) of Mozart's 20th piano concerto, one plays the corresponding compact disc on the hi-fi system. Any image generator is a rudimentary sort of virtual-reality generator, but the term 'virtual reality' is usually reserved for cases where there is both a wide coverage of the user's sensory range, and a substantial element of interaction ('kicking back') between the user and the simulated entities.

Present-day video games do allow interaction between the player and the game objects, but usually only a small fraction of the user's sensory range is covered. The rendered 'environment' consists of images on a small screen, and a proportion of the sounds that the user hears. But virtual-reality video games more worthy of the term do already exist. Typically, the user wears a helmet with built-in headphones and two television screens, one for each eye, and perhaps special gloves and other clothing lined with electrically controlled effectors (pressure-generating devices). There are also sensors that detect the motion of parts of the user's body, especially the head. The information about what the user is doing is passed to a computer, which calculates what the user should be seeing, hearing and feeling, and responds by sending appropriate signals to the image generators (Figure 5.1). When the user looks to the left or right, the pictures on the two television screens pan, just as a real field of view would, to show whatever is on the user's left or right in the simulated world. The user can reach out and pick up a simulated object, and it feels real because the effectors in the glove generate the 'tactile feedback' appropriate to whatever position and orientation the object is seen in.

Game-playing and vehicle simulation are the main uses of virtual reality at present, but a plethora of new uses is envisaged for the

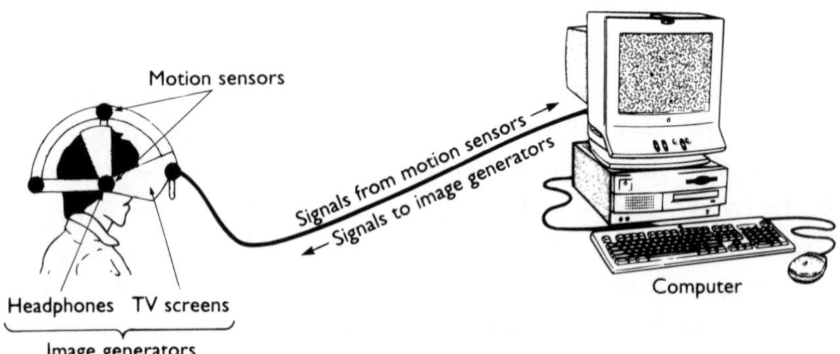

FIGURE 5.1 *Virtual reality as it is implemented today.*

near future. It will soon be commonplace for architects to create virtual-reality prototypes of buildings in which clients can walk around and try out modifications at a stage when they can be implemented relatively effortlessly. Shoppers will be able to walk (or indeed fly) around in virtual-reality supermarkets without ever leaving home, and without ever encountering crowds of other shoppers or listening to music they don't like. Nor will they necessarily be alone in the simulated supermarket, for any number of people can go shopping together in virtual reality, each being provided with images of the others as well as of the supermarket, without any of them having to leave home. Concerts and conferences will be held without venues; not only will there be savings on the cost of the auditorium, and on accommodation and travel, but there is also the benefit that all the participants could be allowed to sit in the best seats simultaneously.

If Bishop Berkeley or the Inquisition had known of virtual reality, they would probably have seized upon it as the perfect illustration of the deceitfulness of the senses, backing up their arguments against scientific reasoning. What would happen if the pilot of a flight simulator tried to use Dr Johnson's test for reality? Although the simulated aircraft and its surroundings do not really exist, they do 'kick back' at the pilot just as they would if they did exist. The pilot can open the throttle and hear the engines roar in response, and feel their thrust through the seat, and see them through the window, vibrating and blasting out hot gas, in spite of the fact that there are no engines there at all. The pilot may experience flying the aircraft through a storm, and hear the thunder and see the rain driving against the windscreen, though none of those things is there in reality. What is outside the cockpit in reality is just a computer, some hydraulic jacks, television screens and loudspeakers, and a perfectly dry and stationary room.

Does this invalidate Dr Johnson's refutation of solipsism? No. His conversation with Boswell could just as well have taken place inside a flight simulator. 'I refute it *thus*', he might have said, opening the throttle and feeling the simulated engine kick back. There is no engine there. What kicks back is ultimately a computer, running a program that calculates what an engine would do if it were 'kicked'. But those calculations, which are external to Dr Johnson's mind, respond to the throttle control in the same complex and autonomous way as the engine would. Therefore they pass the test for reality, and rightly so, for in fact these calculations are physical processes within the computer, and the computer is an ordinary physical object – no less so than an engine – and perfectly real. The fact that it is not a real *engine* is irrelevant to the argument against solipsism. After all, not everything that is real has to be easy to identify. It would not have mattered, in Dr Johnson's original demonstration, if what seemed to be a rock had later turned out to be an animal with a rock-like camouflage, or a holographic projection disguising a garden gnome. So long as its response was complex and autonomous, Dr Johnson would have been right to conclude that it was caused by something real, outside himself, and therefore that reality did not consist of himself alone.

Nevertheless, the feasibility of virtual reality may seem an uncomfortable fact for those of us whose world-view is based on science. Just think what a virtual-reality generator is, from the point of view of physics. It is of course a physical object, obeying the same laws of physics as all other objects do. But it can 'pretend' otherwise. It can pretend to be a completely different object, obeying false laws of physics. Moreover, it can pretend this in a complex and autonomous way. When the user kicks it to test the reality of what it purports to be, it kicks back as if it really were that other, non-existent object, and as if the false laws were true. If we had only such objects to learn physics from, we would learn the wrong laws. (Or would we? Surprisingly, things are not as straightforward as that. I shall return to this question in the next chapter, but first we must consider the phenomenon of virtual reality more carefully.)

On the face of it, Bishop Berkeley would seem to have a point, that virtual reality is a token of the coarseness of human faculties – that its feasibility should warn us of inherent limitations on the capacity of human beings to understand the physical world. Virtual-reality rendering might seem to fall into the same philosophical category as illusions, false trails and coincidences, for these too are phenomena which seem to show us something real but actually mislead us. We have seen that the scientific world-view can accommodate – indeed, expects – the existence of highly misleading phenomena. It is *par excellence* the world-view that can accommodate both human fallibility and external sources of error. Nevertheless, misleading phenomena are basically unwelcome. Except for their curiosity value, or when we learn from them why we are misled, they are things we try to avoid and would rather do without. But virtual reality is not in that category. We shall see that the existence of virtual reality does not indicate that the human capacity to understand the world is inherently limited, but, on the contrary, that it is inherently unlimited. It is no anomaly brought about by the accidental properties of human sense organs, but is a fundamental property of the multiverse at large. And the fact that the multiverse has this property, far from being a minor embarrassment for realism and science, is essential for both – it is the very property that makes science possible. It is not something that 'we would rather do without'; it is something that we literally could not do without.

These may seem rather lofty claims to make on behalf of flight simulators and video games. But it is the phenomenon of virtual reality in general that occupies a central place in the scheme of things, not any particular virtual-reality generator. So I want to consider virtual reality in as general a way as possible. What, if any, are its ultimate limits? What sorts of environment can in principle be artificially rendered, and with what accuracy? By 'in principle' I mean ignoring transient limitations of technology, but taking into account all limitations that may be imposed by the principles of logic and physics.

The way I have defined it, a virtual-reality generator is a machine that gives the user experiences of some real or imagined environment (such as an aircraft) which is, or seems to be, outside the user's mind. Let me call those *external experiences*. External experiences are to be contrasted with *internal experiences* such as one's nervousness when making one's first solo landing, or one's surprise at the sudden appearance of a thunderstorm out of a clear blue sky. A virtual-reality generator indirectly causes the user to have internal experiences as well as external ones, but it cannot be programmed to render a specific internal experience. For example, a pilot who makes roughly the same flight twice in the simulator will have roughly the same external experiences on both occasions, but on the second occasion will probably be less surprised when the thunderstorm appears. Of course on the second occasion the pilot would probably also react differently to the appearance of the thunderstorm, and that would make the subsequent external experiences different too. But the point is that although one can program the machine to make a thunderstorm appear in the pilot's field of view whenever one likes, one cannot program it to make the pilot think whatever one likes in response.

One can conceive of a technology beyond virtual reality, which could also induce specified *internal* experiences. A few internal experiences, such as moods induced by certain drugs, can already be artificially rendered, and no doubt in future it will be possible to extend that repertoire. But a generator of specifiable internal experiences would in general have to be able to override the normal functioning of the user's mind as well as the senses. In other words, it would be replacing the user by a different person. This puts such machines into a different category from virtual-reality generators. They will require quite different technology and will raise quite different philosophical issues, which is why I have excluded them from my definition of virtual reality.

Another type of experience which certainly cannot be artificially rendered is a *logically impossible* one. I have said that a flight simulator can create the experience of a physically impossible flight through a mountain. But nothing can create the experience of factorizing the number 181, because that is logically impossible: 181 is a prime number. (*Believing* that one has factorized 181 is a logically possible experience, but an internal one, and so also outside the scope of virtual reality.) Another logically impossible experience is unconsciousness, for when one is unconscious one is by definition not experiencing anything. Not experiencing anything is quite different from experiencing a total lack of sensations – sensory isolation – which is of course a physically possible environment.

Having excluded logically impossible experiences and internal experiences, we are left with the vast class of *logically possible, external experiences* – experiences of environments which are logically possible, but may or may not be physically possible (Table 5.1). Something is physically possible if it is not forbidden by the laws of physics. In this book I shall assume that the 'laws of physics' include an as yet unknown rule determining the initial state or other supplementary data necessary to give, in principle, a complete description of the multiverse (otherwise these data would be a set of intrinsically inexplicable facts). In that case, an environment is physically possible if and only if it actually exists somewhere in the multiverse (i.e. in some universe or universes). Something is physically impossible if it does not happen anywhere in the multiverse.

I define the *repertoire* of a virtual-reality generator as the set of real or imaginary environments that the generator can be programmed to give the user the experience of. My question about the ultimate limits of virtual reality can be stated like this: what constraints, if any, do the laws of physics impose on the repertoires of virtual-reality generators?

Virtual reality always involves the creation of artificial sense-impressions – image generation – so let us begin there. What

	Logically possible experiences		
	Physically possible environment	Physically impossible environment	Logically impossible experiences
External experiences	e.g. Piloting an aircraft.	e.g. Flying faster than the speed of light.	e.g. Factorizing a prime number.
Internal experiences	e.g. Being proud of one's piloting abilities.	e.g. Experiencing colours outside the visible range.	e.g. Unconsciousness.

TABLE 5.1 *A classification of experiences, with examples of each. Virtual reality is concerned with the generation of logically possible, external experiences (top-left region of the table).*

constraints do the laws of physics impose on the ability of image generators to create artificial images, to render detail and to cover their respective sensory ranges? There are obvious ways in which the detail rendered by a present-day flight simulator could be improved, for example by using higher-definition televisions. But can a realistic aircraft and its surroundings be rendered, even in principle, with the ultimate level of detail – that is, with the greatest level of detail the pilot's senses can resolve? For the sense of hearing, that ultimate level has almost been achieved in hi-fi systems, and for sight it is within reach. But what about the other senses? Is it obvious that it is physically possible to build a general-purpose chemical factory that can produce any specified combination of millions of different odoriferous chemicals at a moment's notice? Or a machine which, when inserted into a gourmet's mouth, can assume the taste and texture of any possible dish – to say nothing of creating the hunger and thirst that precede the meal and the physical satisfaction that follows it? (Hunger and thirst, and other sensations such as balance and muscle tension, are perceived as being internal to the body, but they are external to the mind and are therefore potentially within the scope of virtual reality.)

The difficulty of making such machines may be merely technological, but what about this: suppose that the pilot of a flight simulator aims the simulated aircraft vertically upwards at high speed and then switches off the engines. The aircraft should continue to rise until its upward momentum is exhausted, and then begin to fall back with increasing speed. The whole motion is called *free fall*, even though the aircraft is travelling upwards at first, because it is moving under the influence of gravity alone. When an aircraft is in free fall its occupants are weightless and can float around the cabin like astronauts in orbit. Weight is restored only when an upward force is again exerted on the aircraft, as it soon must be, either by aerodynamics or by the unforgiving ground. (In practice free fall is usually achieved by flying the aircraft under power in the same parabolic trajectory that it would follow in the absence of both engine force and air resistance.) Free-falling aircraft are used to give astronauts weightlessness training before they go into space. A real aircraft could be in free fall for a couple of minutes or more, because it has several kilometres in which to go up and down. But a flight simulator on the ground can be in free fall only for a moment, while its supports let it ride up to their maximum extension and then drop back. Flight simulators (present-day ones, at least) cannot be used for weightlessness training: one needs real aircraft.

Could one remedy this deficiency in flight simulators by giving them the capacity to simulate free fall on the ground (in which case they could also be used as *space*flight simulators)? Not easily, for the laws of physics get in the way. Known physics provides no way other than free fall, even in principle, of removing an object's weight. The only way of putting a flight simulator into free fall while it remained stationary on the surface of the Earth would be somehow to suspend a massive body, such as another planet of similar mass, or a black hole, above it. Even if this were possible (remember, we are concerned here not with immediate practicality, but with what the laws of physics do or do not permit), a real aircraft could also produce frequent, complex changes in the magnitude and direction of the occupants' weight by manoeuvring or by switching its engines on and off. To simulate these changes, the massive body would have to be moved around just as frequently, and it seems likely that the speed of light (if nothing else) would impose an absolute limit on how fast this could be done.

However, to simulate free fall a flight simulator would not have to provide real weightlessness, only the experience of weightlessness, and various techniques which do not involve free fall have been used to approximate that. For example, astronauts train under water in spacesuits that are weighted so as to have zero buoyancy. Another technique is to use a harness that carries the astronaut through the air under computer control to mimic weightlessness. But these methods are crude, and the sensations they produce could hardly be mistaken for the real thing, let alone be indistinguishable from it. One is inevitably supported by forces on one's skin, which one cannot help feeling. Also, the characteristic sensation of falling, experienced through the sense organs in the inner ear, is not rendered at all. One can imagine further improvements: the use of supporting fluids with very low viscosity; drugs that create the sensation of falling. But could one ever render the experience perfectly, in a flight simulator that remained firmly on the ground? If not, then there would be an absolute limit on the fidelity with which flying experiences can ever be rendered artificially. To distinguish between a real aircraft and a simulation, a pilot would only have to fly it in a free-fall trajectory and see whether weightlessness occurred or not.

Stated generally, the problem is this. To override the normal functioning of the sense organs, we must send them images resembling those that would be produced by the environment being simulated. We must also intercept and suppress the images produced by the user's actual environment. But these image manipulations are physical operations, and can be performed only by processes available in the real physical world. Light and sound can be physically absorbed and replaced fairly easily. But as I have said, that is not true of gravity: the laws of physics do not happen to permit it. The example of weightlessness seems to suggest that accurate simulation of a weightless environment by a machine that was not actually in flight might violate the laws of physics.

But that is not so. Weightlessness and all other sensations can, in principle, be rendered artificially. Eventually it will become possible to bypass the sense organs altogether and directly stimulate the nerves that lead from them to the brain.

So, we do not need general-purpose chemical factories or impossible artificial-gravity machines. When we have understood the olfactory organs well enough to crack the code in which they send signals to the brain when they detect scents, a computer with suitable connections to the relevant nerves could send the brain the same signals. Then the brain could experience the scents without the corresponding chemicals ever having existed. Similarly, the brain could experience the authentic sensation of weightlessness even under normal gravity. And of course, no televisions or headphones would be needed either.

Thus the laws of physics impose no limit on the range and

accuracy of image generators. There is no possible sensation, or sequence of sensations, that human beings are capable of experiencing that could not in principle be rendered artificially. One day, as a generalization of movies, there will be what Aldous Huxley in *Brave New World* called 'feelies' – movies for all the senses. One will be able to feel the rocking of a boat beneath one's feet, hear the waves and smell the sea, see the changing colours of the sunset on the horizon and feel the wind in one's hair (whether or not one has any hair) – all without leaving dry land or venturing out of doors. Not only that, feelies will just as easily be able to depict scenes that have never existed, and never could exist. Or they could play the equivalent of music: beautiful abstract combinations of sensations composed to delight the senses.

That every possible sensation can be artificially rendered is one thing; that it will one day be possible, once and for all, to build a single machine that can render any possible sensation calls for something extra: universality. A feelie machine with that capability would be a *universal image generator*.

The possibility of a universal image generator forces us to change our perspective on the question of the ultimate limits of feelie technology. At present, progress in such technology is all about inventing more diverse and more accurate ways of stimulating sense organs. But that class of problems will disappear once we have cracked the codes used by our sense organs, and developed a sufficiently delicate technique for stimulating nerves. Once we can artificially generate nerve signals accurately enough for the brain not to be able to perceive the difference between those signals and the ones that our sense organs would send, increasing the accuracy of this technique will no longer be relevant. At that point the technology will have come of age, and the challenge for further improvement will be not how to render given sensations, but which sensations to render. In a limited domain this is happening today, as the problem of how to get the highest possible fidelity of sound reproduction has come close to being solved with the compact disc and the present generation of sound-reproduction equipment. Soon there will no longer be such a thing as a hi-fi enthusiast. Enthusiasts for sound reproduction will no longer be concerned with how accurate the reproduction is – it will routinely be accurate to the limit of human discrimination – but only with what sounds should be recorded in the first place.

If an image generator is playing a recording taken from life, its *accuracy* may be defined as the closeness of the rendered images to the ones that a person in the original situation would have perceived. More generally, if the generator is rendering artificially designed images, such as a cartoon, or music played from a written composition, the accuracy is the closeness of the rendered images to the intended ones. By 'closeness' we mean closeness as perceived by the user. If the rendering is so close as to be indistinguishable by the user from what is intended, then we can call it *perfectly accurate*. (So a rendering that is perfectly accurate for one user may contain inaccuracies that are perceptible to a user with sharper senses, or with additional senses.)

A universal image generator does not of course contain recordings of all possible images. What makes it universal is that, given a recording of any possible image, it can evoke the corresponding sensation in the user. With a universal auditory sensation generator – the ultimate hi-fi system – the recording might be given in the form of a compact disc. To accommodate auditory sensations that last longer than the disc's storage capacity allows, we must incorporate a mechanism that can feed any number of discs consecutively into the machine. The same proviso holds for all other universal image generators, for strictly speaking an image generator is not universal unless it includes a mechanism for playing recordings of unlimited duration. Furthermore, when the machine has been playing for a long time it will require maintenance, otherwise the images it generates will become degraded or may cease altogether. These and similar considerations are all connected with the fact that considering a single physical object in isolation from the rest of the universe is always an approximation. A universal image generator is universal only in a certain external context, in which it is assumed to be provided with such things as an energy supply, a cooling mechanism and periodic maintenance. That a machine has such external needs does not disqualify it from being regarded as a 'single, universal machine' provided that the laws of physics do not forbid these needs from being met, and provided that meeting those needs does not necessitate changing the machine's design.

Now, as I have said, image generation is only one component of virtual reality: there is the all-important interactive element as well. A virtual-reality generator can be thought of as an image generator whose images are not wholly specified in advance but depend partly on what the user chooses to do. It does not play its user a predetermined sequence of images, as a movie or a feelie would. It composes the images as it goes along, taking into account a continuous stream of information about what the user is doing. Present-day virtual-reality generators, for instance, keep track of the position of the user's head, using motion sensors as shown in Figure 5.1. Ultimately they will have to keep track of everything the user does that could affect the subjective appearance of the simulated environment. The environment may include the user's own body: since the body is external to the mind, the specification of a virtual-reality environment may legitimately include the requirement that the user's body should seem to have been replaced by a new one with specified properties.

The human mind affects the body and the outside world by emitting nerve impulses. Therefore a virtual-reality generator can in principle obtain all the information it needs about what the user is doing by intercepting the nerve signals coming from the user's brain. Those signals, which would have gone to the user's body, can instead be transmitted to a computer and decoded to determine exactly how the user's body would have moved. The signals sent back to the brain by the computer can be the same as those that would have been sent by the body if it were in the specified environment. If the specification called for it, the simulated body could also react differently from the real one, for example to enable it to survive in simulations of environments that would kill a real human body, or to simulate malfunctions of the body.

I had better admit here that it is probably too great an idealization to say that the human mind interacts with the outside world *only* by emitting and receiving nerve impulses. There are chemical messages passing in both directions as well. I am assuming that in principle those messages could also be intercepted and replaced at some point between the brain and the rest of the body. Thus the user would lie motionless, connected to the computer, but having the experience of interacting fully with a simulated world – in effect, living there. Figure 5.2 illustrates what I am envisaging. Incidentally, though such technology lies well in the future, the idea for it is much older than the theory of computation itself. In the early seventeenth century Descartes was already considering the philosophical implications of a sense-manipulating 'demon' that was essentially a virtual-reality generator of the type shown in Figure 5.2, with a supernatural mind replacing the computer.

From the foregoing discussion it seems that any virtual-reality generator must have at least three principal components:

> a set of sensors (which may be nerve-impulse detectors) to detect what the user is doing,
>
> a set of image generators (which may be nerve-stimulation devices), and
>
> a computer in control.

My account so far has concentrated on the first two of these, the sensors and the image generators. That is because, at the present

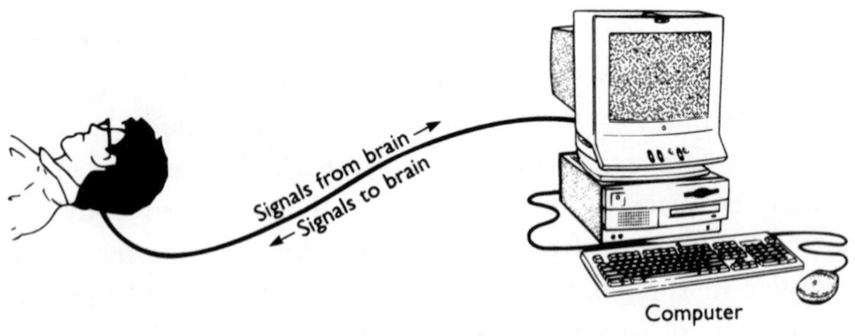

FIGURE 5.2 *Virtual reality as it might be implemented in the future.*

primitive state of the technology, virtual-reality research is still preoccupied with image generation. But when we look beyond transient technological limitations, we see that image generators merely provide the interface – the 'connecting cable' – between the user and the true virtual-reality generator, which is the computer. For it is entirely within the computer that the specified environment is simulated. It is the computer that provides the complex and autonomous 'kicking back' that justifies the word 'reality' in 'virtual reality'. The connecting cable contributes nothing to the user's perceived environment, being from the user's point of view 'transparent', just as we naturally do not perceive our own nerves as being part of our environment. Thus virtual-reality generators of the future would be better described as having only one principal component, a computer, together with some trivial peripheral devices.

I do not want to understate the practical problems involved in intercepting all the nerve signals passing into and out of the human brain, and in cracking the various codes involved. But this is a finite set of problems that we shall have to solve once only. After that, the focus of virtual-reality technology will shift once and for all to the computer, to the problem of programming it to render various environments. What environments we shall be able to render will no longer depend on what sensors and image generators we can build, but on what environments we can specify. 'Specifying' an environment will mean supplying a program for the computer, which is the heart of the virtual-reality generator.

Because of the interactive nature of virtual reality, the concept of an accurate rendering is not as straightforward for virtual reality as it is for image generation. As I have said, the accuracy of an image generator is a measure of the closeness of the rendered images to the intended ones. But in virtual reality there are usually no particular *images* intended: what is intended is a certain environment for the user to experience. Specifying a virtual-reality environment does not mean specifying what the user will experience, but rather specifying how the environment would respond to each of the user's possible actions. For example, in a simulated tennis game one may specify in advance the appearance of the court, the weather, the demeanour of the audience and how well the opponent should play. But one does not specify how the game will go: that depends on the stream of decisions the user makes during the game. Each set of decisions will result in different responses from the simulated environment, and therefore in a different tennis game.

The number of possible tennis games that can be played in a single environment – that is, rendered by a single program – is very large. Consider a rendering of the Centre Court at Wimbledon from the point of view of a player. Suppose, very conservatively, that in each second of the game the player can move in one of two perceptibly different ways (perceptibly, that is, to the player). Then after two seconds there are four possible games, after three seconds, eight possible games, and so on. After about four minutes the number of possible games that are perceptibly different from one another exceeds the number of atoms in the universe, and it continues to rise exponentially. For a program to render that one environment accurately, it must be capable of responding in any one of those myriad, perceptibly different ways, depending on how the player chooses to behave. If two programs respond in the same way to every possible action by the user, then they render the same environment; if they would respond perceptibly differently to even one possible action, they render different environments.

That remains so even if the user never happens to perform the action that shows up the difference. The environment a program renders (for a given type of user, with a given connecting cable) is a logical property of the program, independent of whether the program is ever executed. A rendered environment is accurate in so far as it *would* respond in the intended way to every possible action of the user. Thus its accuracy depends not only on experiences which users of it actually have, but also on experiences they do not have, but would have had if they had chosen to behave differently during the rendering. This may sound paradoxical, but as I have said, it is a straightforward consequence of the fact that virtual reality is, like reality itself, interactive.

This gives rise to an important difference between image

generation and virtual-reality generation. The accuracy of an image generator's rendering can in principle be experienced, measured and certified by the user, but the accuracy of a virtual-reality rendering never can be. For example, if you are a music-lover and know a particular piece well enough, you can listen to a performance of it and confirm that it is a perfectly accurate rendering, in principle down to the last note, phrasing, dynamics and all. But if you are a tennis fan who knows Wimbledon's Centre Court perfectly, you can never confirm that a purported rendering of it is accurate. Even if you are free to explore the rendered Centre Court for however long you like, and to 'kick' it in whatever way you like, and even if you have equal access to the real Centre Court for comparison, you cannot ever certify that the program does indeed render the real location. For you can never know what would have happened if only you had explored a little more, or looked over your shoulder at the right moment. Perhaps if you had sat on the rendered umpire's chair and shouted 'fault!', a nuclear submarine would have surfaced through the grass and torpedoed the scoreboard.

On the other hand, if you find even one difference between the rendering and the intended environment, you can immediately certify that the rendering is *in*accurate. Unless, that is, the rendered environment has some intentionally unpredictable features. For example, a roulette wheel is designed to be unpredictable. If we make a film of roulette being played in a casino, that film may be said to be accurate if the numbers that are shown coming up in the film are the same numbers that actually came up when the film was made. The film will show the same numbers every time it is played: it is totally predictable. So an accurate *image* of an unpredictable environment must be predictable. But what does it mean for a virtual-reality *rendering* of a roulette wheel to be accurate? As before, it means that a user should not find it perceptibly different from the original. But this implies that the rendering must *not* behave identically to the original: if it did, either it or the original could be used to predict the other's behaviour, and then neither would be unpredictable. Nor must it behave in the same way every time it is run. A perfectly rendered roulette wheel must be just as usable for gambling as a real one. Therefore it must be just as unpredictable. Also, it must be just as fair; that is, all the numbers must come up purely randomly, with equal probabilities.

How do we recognize unpredictable environments, and how do we confirm that purportedly random numbers are distributed fairly? We check whether a rendering of a roulette wheel meets its specifications in the same way that we check whether the real thing does: by kicking (spinning) it, and seeing whether it responds as advertised. We make a large number of similar observations and perform statistical tests on the outcomes. Again, however many tests we carry out, we cannot certify that the rendering is accurate, or even that it is probably accurate. For however randomly the numbers seem to come up, they may nevertheless fall into a secret pattern that would allow a user in the know to predict them. Or perhaps if we had asked out loud the date of the battle of Waterloo, the next two numbers that came up would invariably show that date: 18, 15. On the other hand, if the sequence that comes up looks unfair, we cannot know for sure that it is, but we might be able to say that the rendering is *probably* inaccurate. For example, if zero came up on our rendered roulette wheel on ten consecutive spins, we should conclude that we probably do not have an accurate rendering of a fair roulette wheel.

When discussing image generators, I said that the accuracy of a rendered image depends on the sharpness and other attributes of the user's senses. With virtual reality that is the least of our problems. Certainly, a virtual-reality generator that renders a given environment perfectly for humans will not do so for dolphins or extraterrestrials. To render a given environment for a user with given types of sense organs, a virtual-reality generator must be physically adapted to such sense organs and its computer must be programmed with their characteristics. However, the modifications that have to be made to accommodate a given species of user are finite, and need only be carried out once. They amount to what I have called constructing a new 'connecting cable'. As we consider environments of ever greater complexity, the task of rendering environments for a given type of user becomes dominated by writing the programs for calculating what those environments will do; the species-specific part of the task, being of fixed complexity, becomes negligible by comparison. This discussion is about the ultimate limits of virtual reality, so we are considering arbitrarily accurate, long and complex renderings. That is why it makes sense to speak of 'rendering a given environment' without specifying who it is being rendered for.

We have seen that there is a well-defined notion of the accuracy of a virtual-reality rendering: accuracy is the closeness, as far as is perceptible, of the rendered environment to the intended one. But it must be close for every possible way in which the user might behave, and that is why, no matter how observant one is when experiencing a rendered environment, one cannot certify that it is accurate (or probably accurate). But experience can sometimes show that a rendering is inaccurate (or probably inaccurate).

This discussion of accuracy in virtual reality mirrors the relationship between theory and experiment in science. There too, it is possible to confirm experimentally that a general theory is false, but never that it is true. And there too, a short-sighted view of science is that it is all about predicting our sense-impressions. The correct view is that, while sense-impressions always play a role, what science is about is understanding the whole of reality, of which only an infinitesimal proportion is ever experienced.

The program in a virtual-reality generator embodies a general, predictive theory of the behaviour of the rendered environment. The other components deal with keeping track of what the user is doing and with the encoding and decoding of sensory data; these, as I have said, are relatively trivial functions. Thus if the environment is physically possible, rendering it is essentially equivalent to finding rules for predicting the outcome of every experiment that could be performed in that environment. Because of the way in which scientific knowledge is created, ever more accurate predictive rules can be discovered only through ever better explanatory theories. So accurately rendering a physically possible environment depends on understanding its physics.

The converse is also true: discovering the physics of an environment depends on creating a virtual-reality rendering of it. Normally one would say that scientific theories only describe and explain physical objects and processes, but do not render them. For example, an explanation of eclipses of the Sun can be printed in a book. A computer can be programmed with astronomical data and physical laws to predict an eclipse, and to print out a description of it. But rendering the eclipse in virtual reality would require both further programming and further hardware. However, those are already present in our brains! The words and numbers printed by the computer amount to 'descriptions' of an eclipse only because someone knows the meanings of those symbols. That is, the symbols evoke in the reader's mind some sort of likeness of some predicted effect of the eclipse, against which the real appearance of that effect will be tested. Moreover, the 'likeness' that is evoked is interactive. One can observe an eclipse in many ways: with the naked eye, or by photography, or using various scientific instruments; from some positions on Earth one will see a total eclipse of the Sun, from other positions a partial eclipse, and from anywhere else no eclipse at all. In each case an observer will experience different images, any of which can be predicted by the theory. What the computer's description evokes in a reader's mind is not just a single image or sequence of images, but a general method of creating many different images, corresponding to the many ways in which the reader may contemplate making observations. In other words, it is a virtual-reality rendering. Thus, in a broad enough sense, taking into account the processes that must take place inside the scientist's mind, science and the virtual-reality rendering of physically possible environments are two terms denoting the same activity.

Now, what about the rendering of environments that are not physically possible? On the face of it, there are two distinct types of virtual-reality rendering: a minority that depict physically possible environments, and a majority that depict physically impossible environments. But can this distinction survive closer examination? Consider a virtual-reality generator in the act of rendering a physically impossible environment. It might be a flight simulator, running a program that calculates the view from the cockpit of an aircraft that can fly faster than light. The flight simulator is *rendering* that environment. But in addition the flight simulator is itself the environment that the user is experiencing, in the sense that it is a physical object surrounding the user. Let us consider this environment. Clearly it is a physically possible environment. Is it a renderable environment? Of course. In fact it is exceptionally easy to render: one simply uses a second flight simulator of the same design, running the identical program. Under those circumstances the second flight simulator can be thought of as rendering either the physically impossible aircraft, or a physically possible environment, namely the first flight simulator. Similarly, the first flight simulator could be regarded as rendering a physically possible environment, namely the second flight simulator. If we assume that any virtual-reality generator that can in principle be built, can in principle be built again, then it follows that every virtual-reality generator, running any program in its repertoire, is rendering *some* physically possible environment. It may be rendering other things as well, including physically impossible environments, but in particular there is always some physically possible environment that it is rendering.

So, which physically impossible environments can be rendered in virtual reality? Precisely those that are not perceptibly different from physically possible environments. Therefore the connection between the physical world and the worlds that are renderable in virtual reality is far closer than it looks. We think of some virtual-reality renderings as depicting fact, and others as depicting fiction, but the fiction is always an interpretation in the mind of the beholder. There is no such thing as a virtual-reality environment that the user would be compelled to interpret as physically impossible.

We might choose to render an environment as predicted by some 'laws of physics' that are different from the true laws of physics. We may do this as an exercise, or for fun, or as an approximation because the true rendering is too difficult or expensive. If the laws we are using are as close as we can make them to real ones, given the constraints under which we are operating, we may call these renderings 'applied mathematics' or 'computing'. If the rendered objects are very different from physically possible ones, we may call the rendering 'pure mathematics'. If a physically impossible environment is rendered for fun, we call it a 'video game' or 'computer art'. All these are interpretations. They may be useful interpretations, or even essential in explaining our motives in composing a particular rendering. But as far as the rendering itself goes there is always an alternative interpretation, namely that it accurately depicts some physically possible environment.

It is not customary to think of mathematics as being a form of virtual reality. We usually think of mathematics as being about abstract entities, such as numbers and sets, which do not affect the senses; and it might therefore seem that there can be no question of artificially rendering their effect on us. However, although mathematical entities do not affect the senses, the experience of doing mathematics is an external experience, no less than the experience of doing physics is. We make marks on pieces of paper and look at them, or we imagine looking at such marks – indeed, we cannot do mathematics without imagining abstract mathematical entities. But this means imagining an environment whose 'physics' embodies the complex and autonomous properties of those entities. For example, when we imagine the abstract concept of a line segment which has no thickness, we may imagine a line that is visible but imperceptibly wide. That much may, just about, be arranged in physical reality. But mathematically the line must continue to have no thickness when we view it under arbitrarily powerful magnification. That is not a property of any physical line, but it can easily be achieved in the virtual reality of our imagination.

Imagination is a straightforward form of virtual reality. What may not be so obvious is that our 'direct' experience of the world through our senses is virtual reality too. For our external experience is never direct; nor do we even experience the signals in our nerves directly – we would not know what to make of the streams of electrical crackles that they carry. What we experience directly is a virtual-reality rendering, conveniently generated for us by our unconscious minds from sensory data plus complex inborn and acquired theories (i.e. programs) about how to interpret them.

We realists take the view that reality is out there: objective, physical and independent of what we believe about it. But we never experience that reality directly. Every last scrap of our external experience is of virtual reality. And every last scrap of our knowledge – including our knowledge of the non-physical worlds of logic, mathematics and philosophy, and of imagination, fiction, art and fantasy – is encoded in the form of programs for the rendering of those worlds on our brain's own virtual-reality generator.

So it is not just science – reasoning about the physical world – that involves virtual reality. All reasoning, all thinking and all external experience are forms of virtual reality. These things are physical processes which so far have been observed in only one place in the universe, namely the vicinity of the planet Earth. We shall see in Chapter 8 that all living processes involve virtual reality too, but human beings in particular have a special relationship with it. Biologically speaking, the virtual-reality rendering of their environment is the characteristic means by which human beings survive. In other words, it is the reason why human beings exist. The ecological niche that human beings occupy depends on virtual reality as directly and as absolutely as the ecological niche that koala bears occupy depends on eucalyptus leaves.

TERMINOLOGY

image generator A device that can generate specifiable sensations for a user.

universal image generator An image generator that can be programmed to generate any sensation that the user is capable of experiencing.

external experience An experience of something outside one's own mind.

internal experience An experience of something within one's own mind.

physically possible Not forbidden by the laws of physics. An environment is physically possible if and only if it exists somewhere in the multiverse (on the assumption that the initial conditions and all other supplementary data of the multiverse are determined by some as yet unknown laws of physics).

logically possible Self-consistent.

virtual reality Any situation in which the user is given the experience of being in a specified environment.

repertoire The repertoire of a virtual-reality generator is the set of environments that the generator can be programmed to give the user the experience of.

image Something that gives rise to sensations.

accuracy An image is accurate in so far as the sensations it generates are close to the intended sensations.

A rendered environment is accurate in so far as it would respond in the intended way to every possible action of the user.

perfect accuracy Accuracy so great that the user cannot distinguish the image or rendered environment from the intended one.

SUMMARY

Virtual reality is not just a technology in which computers simulate the behaviour of physical environments. The fact that virtual reality is possible is an important fact about the fabric of reality. It is the basis not only of computation, but of human imagination and external experience, science and mathematics, art and fiction.

What are the ultimate limits – the full scope – of virtual reality (and hence of computation, science, imagination and the rest)? In the next chapter we shall see that in one respect the scope of virtual reality is unlimited, while in another it is drastically circumscribed.

6

Universality and the Limits of Computation

The heart of a virtual-reality generator is its computer, and the question of what environments can be rendered in virtual reality must eventually come down to the question of what computations can be performed. Even today, the repertoire of virtual-reality generators is limited as much by their computers as by their image generators. Whenever a new, faster computer, with more memory and better image-processing hardware, is incorporated into a virtual-reality generator, the repertoire is enlarged. But will this always be so, or will we eventually encounter full universality, as I have argued we should expect to in the case of image generators? In other words, is there a single virtual-reality generator, buildable once and for all, that could be programmed to render any environment that the human mind is capable of experiencing?

Just as with image generators, we do not mean by this that a single virtual-reality generator might contain within itself the specifications of all logically possible environments. We mean only that for every logically possible environment it would be possible to program the generator to render that environment. We can envisage encoding the programs on, for example, magnetic disks. The more complex the environment, the more disks may be needed to store the corresponding program. So to render complex environments the machine must have a mechanism, just as I have described for the universal image generator, that can read unlimited numbers of disks. Unlike an image generator, a virtual-reality generator may need a growing amount of 'working memory' to store the intermediate results of its calculations. We may envisage this as being provided in the form of blank disks. Once again, the fact that a machine needs to be supplied with energy, blank disks and maintenance does not prevent us from regarding it as a 'single machine', provided that these operations are not tantamount to changing the machine's design, and are not forbidden by the laws of physics.

In this sense, then, a computer with an effectively unlimited memory capacity can be envisaged in principle. But a computer with an unlimited speed of computation cannot. A computer of given design will always have a fixed maximum speed, which only design changes can increase. Therefore a given virtual-reality generator will not be able to perform unlimited amounts of computation per unit time. Will this not limit its repertoire? If an environment is so complex that the computation of what the user should be seeing one second from now takes the machine more than one second to compute, how can the machine possibly render that environment accurately? To achieve universality, we need a further technological trick.

To extend its repertoire as far as is physically possible, a virtual-reality generator would have to take control of one further attribute of the user's sensory system, namely the processing speed of the user's brain. If the human brain were like an electronic computer, this would simply be a matter of changing the rate at which its 'clock' emits synchronizing pulses. No doubt the brain's 'clock' will not be so easily controlled. But again this presents no problem of principle. The brain is a finite physical object, and all its functions are physical processes which in principle can be slowed down or stopped. The ultimate virtual-reality generator would have to be capable of doing that.

To achieve a perfect rendering of environments which call for a lot of computation, a virtual-reality generator would have to operate in something like the following way. Each sensory nerve is physically capable of relaying signals at a certain maximum rate, because a nerve cell which has fired cannot fire again until about one millisecond later. Therefore, immediately after a particular nerve has fired, the computer has at least one millisecond to decide whether, and when, that nerve should fire again. If it has computed that decision within, say, half a millisecond, no tampering with the brain's speed is necessary, and the computer merely fires the nerve at the appropriate times. Otherwise, the computer causes the brain to slow down (or, if necessary, to stop) until the calculation of what should happen next is complete; it then restores the brain's normal speed. What would this feel like to the user? By definition, like nothing. The user would experience only the environment specified in the program, without any slowing down, stopping or restarting. Fortunately it is never necessary for a virtual-reality generator to make the brain operate *faster* than normal; that would eventually raise problems of principle because, among other things, no signal can travel faster than the speed of light.

This method allows us to specify in advance an arbitrarily complicated environment whose simulation requires any finite amount of computation, and to experience that environment at any subjective speed and level of detail that our minds are capable of assimilating. If the requisite calculations are too numerous for the computer to perform within the subjectively perceived time, the experience will be unaffected, but the user will pay for its complexity in terms of externally elapsed time. The user might emerge from the virtual-reality generator after what seemed subjectively like a five-minute experience, to find that years had passed in physical reality.

A user whose brain is switched off, for however long, and then switched on again will have an uninterrupted experience of some environment. But a user whose brain is switched off for ever has no experiences at all from that moment on. This means that a program which might at some point switch the user's brain off, and never switch it on again, does not generate an environment for the user to experience and therefore does not qualify as a valid program for a virtual-reality generator. But a program which, eventually, always switches the user's brain back on causes the virtual-reality generator to render some environment. Even a program which emits no nerve signals at all renders the dark, silent environment of perfect sensory isolation.

In our search for the ultimate in virtual-reality we have strayed a very long way from what is feasible today, or even from what is on any foreseeable technological horizon. So let me stress again that for our present purposes technological obstacles are irrelevant. We are not investigating what sorts of virtual-reality generator can be built, or even, necessarily, what sorts of virtual-reality generator will ever be built, by human engineers. We are investigating what the laws of physics do and do not allow in the way of virtual reality. The reason why this is important has nothing to do with the prospects for making better virtual-reality generators. It is that the relationship between virtual reality and 'ordinary' reality is part of the deep, unexpected structure of the world, which this book is about.

By considering various tricks – nerve stimulation, stopping and starting the brain, and so on – we have managed to envisage a physically possible virtual-reality generator whose repertoire covers the entire sensory range, is fully interactive, and is not constrained by the speed or memory capacity of its computer. Is there anything outside the repertoire of such a virtual-reality generator? Would its repertoire be the set of all logically possible environments? It would not. Even this futuristic machine's repertoire is drastically circumscribed by the mere fact of its being a physical object. It does not even scratch the surface of what is logically possible, as I shall now show.

The basic idea of the proof – known as a *diagonal argument* – predates the idea of virtual reality. It was first used by the nineteenth-century mathematician Georg Cantor to prove that there are infinite quantities greater than the infinity of natural numbers (1, 2, 3 . . .). The same form of proof is at the heart of the modern theory of computation developed by Alan Turing and others in the 1930s. It was also used by Kurt Gödel to prove his celebrated 'incompleteness theorem', of which more in Chapter 10.

Each environment in our machine's repertoire is generated by some program for its computer. Imagine the set of all valid programs for this computer. From a physical point of view, each such program specifies a particular set of values for physical variables, on the disks or other media, that represent the computer's program.

We know from quantum theory that all such variables are quantized, and therefore that, no matter how the computer works, the set of possible programs is discrete. Each program can therefore be expressed as a finite sequence of symbols in a discrete code or computer language. There are infinitely many such programs, but each one can contain only a finite number of symbols. That is because symbols are physical objects, made of matter in recognizable configurations, and one could not manufacture an infinite number of them. As I shall explain in Chapter 10, these intuitively obvious physical requirements – that the programs must be quantized, and that each of them must consist of a finite number of symbols and can be executed in a sequence of steps – are more substantive than they seem. They are the only consequences of the laws of physics that are needed as input for the proof, but they are enough to impose drastic restrictions on the repertoire of any physically possible machine. Other physical laws may impose even more restrictions, but they would not affect the conclusions of this chapter.

Now let us imagine this infinite set of possible programs arranged in an infinitely long list, and numbered Program 1, Program 2, and so on. They could, for instance, be arranged in 'alphabetical' order with respect to the symbols in which they are expressed. Because each program generates an environment, this list can also be regarded as a list of all the environments in the machine's repertoire; we may call them Environment 1, Environment 2, and so on. It could be that some of the environments are repeated in the list, because two different programs might in effect perform the same calculations, but that will not affect the argument. What is important is that each environment in our machine's repertoire should appear at least once in the list.

A simulated environment may be limited or unlimited in apparent physical size and apparent duration. An architect's simulation of a house, for example, can be run for an unlimited time, but will probably cover only a limited volume. A video game might allow the user only a finite time for play before the game ends, or it might render a game-universe of unlimited size, allow an unlimited amount of exploration and end only when the user deliberately ends it. To make the proof simpler, let us consider only programs that continue to run for ever. That is not much of a restriction, because if a program halts we can always choose to regard its lack of response as being the response of a sensory-isolation environment.

Let me define a class of logically possible environments which I shall call *Cantgotu environments*, partly in honour of *Cant*or, *Gö*del and *Tu*ring, and partly for a reason I shall explain shortly. They are defined as follows. For the first subjective minute, a Cantgotu environment behaves differently from Environment 1 (generated by Program 1 of our generator). It does not matter how it does behave, so long as it is, to the user, recognizably different from Environment 1. During the second minute it behaves differently from Environment 2 (though it is now allowed to resemble Environment 1 again). During the third minute, it behaves differently from Environment 3, and so on. Any environment that satisfies these rules I shall call a Cantgotu environment.

Now, since a Cantgotu environment does not behave exactly like Environment 1, it cannot *be* Environment 1; since it does not behave exactly like Environment 2, it cannot *be* Environment 2. Since it is guaranteed sooner or later to behave differently from Environment 3, Environment 4 and every other environment on the list, it cannot be any of those either. But that list contains all the environments that are generated by every possible program for this machine. It follows that none of the Cantgotu environments are in the machine's repertoire. The Cantgotu environments are environments that we *can't go to* using this virtual-reality generator.

Clearly there are enormously many Cantgotu environments, because the definition leaves enormous freedom in choosing how they should behave, the only constraint being that during each minute they should not behave in one particular way. It can be proved that, for every environment in the repertoire of a given virtual-reality generator, there are infinitely many Cantgotu environments that it cannot render. Nor is there much scope for extending the repertoire by using a range of different virtual-reality generators. Suppose that we had a hundred of them, each (for the sake of argument) with a different repertoire. Then the whole collection, combined with the programmable control system that determines which of them shall be used to run a given program, is just a larger virtual-reality generator. That generator is subject to the argument I have given, so for every environment it can render there will be infinitely many that it cannot. Furthermore, the assumption that different virtual-reality generators might have different repertoires turns out to be over-optimistic. As we shall see in a moment, all sufficiently sophisticated virtual-reality generators have essentially the same repertoire.

Thus our hypothetical project of building the ultimate virtual-reality generator, which had been going so well, has suddenly run into a brick wall. Whatever improvements may be made in the distant future, the repertoire of the entire technology of virtual reality will never grow beyond a certain fixed set of environments. Admittedly this set is infinitely large, and very diverse by comparison with human experience prior to virtual-reality technology. Nevertheless, it is only an infinitesimal fraction of the set of all logically possible environments.

What would it feel like to be in a Cantgotu environment? Although the laws of physics do not permit us to be in one, it is still logically possible and so it is legitimate to ask what it would feel like. Certainly, it could give us no new *sensations*, because a universal image generator *is* possible and is assumed to be part of our high-technology virtual-reality generator. So a Cantgotu environment would seem mysterious to us only after we had experienced it and reflected on the results. It would go something like this. Suppose you are a virtual-reality buff in the distant, ultra-high-technology future. You have become jaded, for it seems to you that you have already tried everything interesting. But then one day a genie appears and claims to be able to transport you to a Cantgotu environment. You are sceptical, but agree to put its claim to the test. You are whisked away to the environment. After a few experiments you seem to recognize it – it responds just like one of your favourite environments, which on your home virtual-reality system has program number *X*. However, you keep experimenting, and eventually, during the *X*th subjective minute of the experience, the environment responds in a way that is markedly different from anything that Environment *X* would do. So you give up the idea that this is Environment *X*. You may then notice that everything that has happened so far is also consistent with another renderable environment, Environment *Y*. But then, during the *Y*th subjective minute you are proved wrong again. The characteristic of a Cantgotu environment is simply this: no matter how often you guess, no matter how complex a program you contemplate as being the one that might be rendering the environment, you will always be proved wrong because *no* program will render it, on your virtual-reality generator or on any other.

Sooner or later you will have to bring the test to a close. At that point you may well decide to concede the genie's claim. That is not to say that you could ever *prove* that you had been in a Cantgotu environment, for there is always an even more complex program that the genie might have been running, which would match your experiences so far. That is just the general feature of virtual reality that I have already discussed, namely that experience cannot prove that one is in a given environment, be it the Centre Court at Wimbledon or an environment of the Cantgotu type.

Anyway, there are no such genies, and no such environments. So we must conclude that physics does not allow the repertoire of a virtual-reality generator to be anywhere near as large as logic alone would allow. How large can it be?

Since we cannot hope to render all logically possible environments, let us consider a weaker (but ultimately more interesting) sort of universality. Let us define a *universal virtual-reality generator* as one whose repertoire contains that of every other physically possible virtual-reality generator. Can such a machine exist? It can. Thinking about futuristic devices based on computer-controlled nerve stimulation makes this obvious – in fact, almost too obvious. Such a machine could be programmed to have the characteristics of any rival machine. It could calculate how that machine would respond, under any given program, to any behaviour by the user, and so could render those responses with perfect accuracy (from the point of view of any given user). I say that this is 'almost too obvious' because it contains an important assumption about what the proposed device, and more specifically its computer, could be programmed to do: given the appropriate program, and enough time and storage media, it could calculate the output of any computation performed by any other computer, including the one in the rival virtual-reality generator. Thus the feasibility of a universal virtual-reality generator depends on the existence of a universal computer – a single machine that can calculate anything that can be calculated.

As I have said, this sort of universality was first studied not by physicists but by mathematicians. They were trying to make precise the intuitive notion of 'computing' (or 'calculating' or 'proving') something in mathematics. They did not take on board the fact that mathematical calculation is a physical process (in particular, as I have explained, it is a virtual-reality rendering process), so it is impossible to determine by mathematical reasoning what can or cannot be calculated mathematically. That depends entirely on the laws of physics. But instead of trying to deduce their results from physical laws, mathematicians postulated abstract models of 'computation', and *defined* 'calculation' and 'proof' in terms of those models. (I shall discuss this interesting mistake in Chapter 10.) That is how it came about that over a period of a few months in 1936, three mathematicians, Emil Post, Alonzo Church and, most importantly, Alan Turing, independently created the first abstract designs for universal computers. Each of them conjectured that his model of 'computation' did indeed correctly formalize the traditional, intuitive notion of mathematical 'computation'. Consequently, each of them also conjectured that his model was equivalent to (had the same repertoire as) any other reasonable formalization of the same intuition. This is now known as the *Church–Turing conjecture*.

Turing's model of computation, and his conception of the nature of the problem he was solving, was the closest to being physical. His abstract computer, the *Turing machine*, was abstracted from the idea of a paper tape divided into squares, with one of a finite number of easily distinguishable symbols written on each square. Computation was performed by examining one square at a time, moving the tape backwards or forwards, and erasing or writing one of the symbols according to simple, unambiguous rules. Turing proved that one particular computer of this type, the *universal Turing machine*, had the combined repertoire of all other Turing machines. He conjectured that this repertoire consisted precisely of 'every function that would naturally be regarded as computable'. He meant computable *by mathematicians*.

But mathematicians are rather untypical physical objects. Why should we assume that rendering them in the act of performing calculations is the ultimate in computational tasks? It turns out that it is not. As I shall explain in Chapter 9, *quantum computers* can perform computations of which no (human) mathematician will ever, even in principle, be capable. It is implicit in Turing's work that he expected what 'would naturally be regarded as computable' to be also what could, at least in principle, be computed in nature. This expectation is tantamount to a stronger, physical version of the Church–Turing conjecture. The mathematician Roger Penrose has suggested that it should be called the *Turing principle*:

The Turing principle

(for abstract computers simulating physical objects)

There exists an abstract universal computer whose repertoire includes any computation that any physically possible object can perform.

Turing believed that the 'universal computer' in question was the universal Turing machine. To take account of the wider repertoire of quantum computers, I have stated the principle in a form that does not specify which particular 'abstract computer' does the job.

The proof I have given of the existence of Cantgotu environments is essentially due to Turing. As I said, he was not thinking explicitly in terms of virtual reality, but an 'environment that can be rendered' does correspond to a class of mathematical questions whose answers can be calculated. Those questions are *computable*. The remainder, the questions for which there is no way of calculating the answer, are called *non-computable.* If a question is non-computable that does not mean that it has no answer, or that its answer is in any sense ill-defined or ambiguous. On the contrary, it means that it definitely has an answer. It is just that physically there is no way, even in principle, of obtaining that answer (or more precisely, since one could always make a lucky, unverifiable guess, of proving that it is the answer). For example, a *prime pair* is a pair of prime numbers whose difference is 2, such as 3 and 5, or 11 and 13. Mathematicians have tried in vain to answer the question whether there are infinitely many such pairs, or only a finite number of them. It is not even known whether this question is computable. Let us suppose that it is not. That is to say that no one, and no computer, can ever produce a proof either that there are only finitely many prime pairs or that there are infinitely many. Even so, the question does have an answer: one can say with certainty that either there is a highest prime pair or there are infinitely many prime pairs; there is no third possibility. The question remains well-defined, even though we may never know the answer.

In virtual-reality terms: no physically possible virtual-reality generator can render an environment in which answers to non-computable questions are provided to the user on demand. Such environments are of the Cantgotu type. And conversely, every Cantgotu environment corresponds to a class of mathematical questions ('what would happen next in an environment defined in such-and-such a way?') which it is physically impossible to answer.

Although non-computable questions are infinitely more numerous than computable ones, they tend to be more esoteric. That is no accident. It is because the parts of mathematics that we tend to consider the least esoteric are those we see reflected in the behaviour of physical objects in familiar situations. In such cases we can often use those physical objects to answer questions about the corresponding mathematical relationships. For example, we can count on our fingers because the physics of fingers naturally mimics the arithmetic of the whole numbers from zero to ten.

The repertoires of the three very different abstract computers defined by Turing, Church and Post were soon proved to be identical. So have the repertoires of all abstract models of mathematical computation that have since been proposed. This is deemed to lend support to the Church–Turing conjecture and to the universality of the universal Turing machine. However, the computing power of *abstract* machines has no bearing on what is computable in reality. The scope of virtual reality, and its wider implications for the comprehensibility of nature and other aspects of the fabric of reality, depends on whether the relevant computers are physically realizable. In particular, any genuine universal computer must itself be physically realizable. This leads to a stronger version of the Turing principle:

The Turing principle

(for physical computers simulating each other)

It is possible to build a universal computer: a machine that can be programmed to perform any computation that any other physical object can perform.

It follows that if a universal image generator were controlled by a universal computer, the resulting machine would be a universal virtual-reality generator. In other words, the following principle also holds:

The Turing principle

(for virtual-reality generators rendering each other)

It is possible to build a virtual-reality generator whose repertoire includes that of every other physically possible virtual-reality generator.

Now, any environment can be rendered by a virtual-reality generator of *some* sort (for instance, one could always regard a copy of that very environment as a virtual-reality generator with perhaps a very small repertoire). So it also follows from this version of the Turing principle that any physically possible environment can be rendered by the universal virtual-reality generator. Hence to express the very strong self-similarity that exists in the structure of reality, embracing not only computations but all physical processes, the Turing principle can be stated in this all-embracing form:

The Turing principle

It is possible to build a virtual-reality generator whose repertoire includes every physically possible environment.

This is the strongest form of the Turing principle. It not only tells us that various parts of reality can resemble one another. It tells us that a single physical object, buildable once and for all (apart from maintenance and a supply of additional memory when needed), can perform with unlimited accuracy the task of describing or mimicking any other part of the multiverse. The set of all behaviours and responses of that one object exactly mirrors the set of all behaviours and responses of all other physically possible objects and processes.

This is just the sort of self-similarity that is necessary if, according to the hope I expressed in Chapter 1, the fabric of reality is to be truly unified and comprehensible. If the laws of physics as they apply to any physical object or process are to be comprehensible, they must be capable of being embodied in another physical object – the knower. It is also necessary that processes capable of creating such knowledge be physically possible. Such processes are called science. Science depends on experimental testing, which means physically rendering a law's predictions and comparing it with (a rendering of) reality. It also depends on explanation, and that requires the abstract laws themselves, not merely their predictive content, to be capable of being rendered in virtual reality. This is a tall order, but reality does meet it. That is to say, the laws of physics meet it. The laws of physics, by conforming to the Turing principle, make it physically possible for those same laws to become known to physical objects. Thus, the laws of physics may be said to mandate their own comprehensibility.

Since building a universal virtual-reality generator is physically possible, it must actually *be* built in some universes. A caveat is necessary here. As I explained in Chapter 3, we can normally define a physically possible process as one that actually occurs somewhere in the multiverse. But strictly speaking, a universal virtual-reality generator is a limiting case that requires arbitrarily large resources to operate. So what we really mean by saying that it is 'physically possible' is that virtual-reality generators with repertoires arbitrarily close to the set of all physically possible environments exist in the multiverse. Similarly, since the laws of physics are capable of being rendered, they *are* rendered somewhere. Thus it follows from the Turing principle (in the strong form for which I have argued) that the laws of physics do not merely mandate their own comprehensibility in some abstract sense – comprehensibility by abstract scientists, as it were. They imply the physical existence, somewhere in the multiverse, of entities that understand them arbitrarily well. I shall discuss this implication further in later chapters.

Now I return to the question I posed in the previous chapter, namely whether, if we had only a virtual-reality rendering based on the wrong laws of physics to learn from, we should expect to learn the wrong laws. The first thing to stress is that we *do* have only virtual reality based on the wrong laws to learn from! As I have said, all our external experiences are of virtual reality, generated by our own brains. And since our concepts and theories (whether inborn or learned) are never perfect, all our renderings are indeed inaccurate. That is to say, they give us the experience of an environment that is significantly different from the environment that we are really in. Mirages and other optical illusions are examples of this. Another is that we experience the Earth to be at rest beneath our feet, despite its rapid and complex motion in reality. Another is that we experience a single universe, and a single instance of our own conscious selves at a time, while in reality there are many. But these inaccurate and misleading experiences provide no argument against scientific reasoning. On the contrary, such deficiencies are its very starting-point.

We are embarked upon solving problems about physical reality. If it turns out that all this time we have merely been studying the programming of a cosmic planetarium, then that would merely mean that we have been studying a smaller portion of reality than we thought. So what? Such things have happened many times in the history of science, as our horizons have expanded beyond the Earth to include the solar system, our Galaxy, other galaxies, clusters of galaxies and so on, and, of course, parallel universes. Another such broadening may happen tomorrow; indeed, it may happen according to any one of an infinity of possible theories – or it may never happen. Logically, we must concede to solipsism and related doctrines that the reality we are learning about *might* be an unrepresentative portion of a larger, inaccessible or incomprehensible structure. But the general refutation that I have given of such doctrines shows us that it is irrational to build upon that possibility. Following Occam, we shall entertain such theories when, and only when, they provide better explanations than simpler rival theories.

However, there is a question we can still ask. Suppose that someone were imprisoned in a small, unrepresentative portion of our own reality – for instance, inside a universal virtual-reality generator that was programmed with the wrong laws of physics. What could such prisoners learn about our external reality? At first sight, it seems impossible that they could discover anything at all about it. It may seem that the most they could discover would be the laws of operation, i.e. the program, of the computer that operated their prison.

But that is not so! Again, we must bear in mind that if the prisoners are scientists, they will be seeking explanations as well as predictions. In other words, they will not be content with merely knowing the program that operates their prison: they will want to explain the origin and attributes of the various entities, including themselves, that they observe in the reality they inhabit. But in most virtual-reality environments no such explanation exists, for the rendered objects do not originate there but have been designed in the external reality. Suppose that you are playing a virtual-reality video game. For the sake of simplicity, suppose that the game is essentially chess (a first-person-perspective version perhaps, in which you adopt the persona of the king). You will use the normal methods of science to discover this environment's 'laws of physics' and their emergent consequences. You will learn that checkmate and stalemate are 'physically' possible events (i.e. possible under your best understanding of how the environment works), but that a position with nine white pawns is not 'physically' possible. Once you had understood the laws sufficiently well, you would notice that the chessboard is too simple an object to have, for instance, thoughts, and consequently that your own thought-processes cannot be governed by the laws of chess alone. Similarly, you could tell that during any number of games of chess the pieces can never evolve into self-reproducing configurations. And if life cannot evolve on the chessboard, far less can intelligence evolve. Therefore you would also infer that your own thought-processes could not have originated in the universe in which you found yourself. So even if you had lived within the rendered environment all your life, and did not have your own memories of the outside world to account for as well, your knowledge would not be confined to that environment. You would know that, even though the universe seemed to have a certain layout and obey certain laws, there must be a wider universe outside it, obeying different laws of physics. And you could even guess some of the ways in which these wider laws would have to differ from the chessboard laws.

Arthur C. Clarke once remarked that 'any sufficiently advanced technology is indistinguishable from magic'. This is true, but slightly misleading. It is stated from the point of view of a prescientific thinker, which is the wrong way round. The fact is that to anyone who understands what virtual reality is, even genuine magic would be indistinguishable from technology, for there is no room for magic in a comprehensible reality. Anything that seems incomprehensible is regarded by science merely as evidence that there is something we have not yet understood, be it a conjuring trick, advanced technology or a new law of physics.

Reasoning from the premise of one's own existence is called 'anthropic' reasoning. Although it has some applicability in cosmology, it usually has to be supplemented by substantive assumptions about the nature of 'oneself' before it yields definite conclusions. But anthropic reasoning is not the only way in which the inmates of our hypothetical virtual-reality prison could gain knowledge of an outside world. *Any* of their evolving explanations of their narrow world could, at the drop of a hat, reach into an outside reality. For instance, the very rules of chess contain what a thoughtful player may realize is 'fossil evidence' of those rules having had an evolutionary history: there are 'exceptional' moves such as castling and capturing *en passant* which increase the complexity of the rules but improve the game. In explaining that complexity, one justifiably concludes that the rules of chess were not always as they are now.

In the Popperian scheme of things, explanations always lead to new problems which in turn require further explanations. If the prisoners fail, after a while, to improve upon their existing explanations, they may of course give up, perhaps falsely concluding that there are no explanations available. But if they do not give up they will be thinking about those aspects of their environment that seem inadequately explained. Thus if the high-technology jailers wanted to be confident that their rendered environment would forever fool their prisoners into thinking that there is no outside world, they would have their work cut out for them. The longer they wanted the illusion to last, the more ingenious the program would have to be. It is not enough that the inmates be prevented from observing the outside. The rendered environment would also have to be such that no explanations of anything inside would ever require one to postulate an outside. The environment, in other words, would have to be self-contained as regards explanations. But I doubt that any part of reality, short of the whole thing, has that property.

TERMINOLOGY

universal virtual-reality generator One whose repertoire contains every physically possible environment.

Cantgotu environments Logically possible environments which cannot be rendered by any physically possible virtual-reality generator.

diagonal argument A form of proof in which one imagines listing a set of entities, and then uses the list to construct a related entity that cannot be on the list.

Turing machine One of the first abstract models of computation.

universal Turing machine A Turing machine with the combined repertoire of all other Turing machines.

Turing principle (in its strongest form) It is physically possible to build a universal virtual-reality generator.

On the assumptions I have been making, this implies that there is no upper bound on the universality of virtual-reality generators that will actually be built somewhere in the multiverse.

SUMMARY

The diagonal argument shows that the overwhelming majority of logically possible environments cannot be rendered in virtual reality. I have called them Cantgotu environments. There is nevertheless a comprehensive self-similarity in physical reality that is expressed in the Turing principle: *it is possible to build a virtual-reality generator whose repertoire includes every physically possible environment.* So a single, buildable physical object can mimic all the behaviours and responses of any other physically possible object or process. This is what makes reality comprehensible.

It also makes possible the evolution of living organisms. However, before I discuss the theory of evolution, the fourth strand of explanation of the fabric of reality, I must make a brief excursion into epistemology.

7

A Conversation About Justification

(or 'David and the Crypto-inductivist')

I think that I have solved a major philosophical problem: the problem of induction. Karl Popper

As I explained in the Preface, this book is not primarily a defence of the fundamental theories of the four main strands; it is an investigation of what those theories say, and what sort of reality they describe. That is why I do not address opposing theories in any depth. However, there is one opposing theory – namely, common sense – which reason requires me to refute in detail wherever it seems to conflict with what I am asserting. Hence in Chapter 2 I presented a root-and-branch refutation of the common-sense idea that there is only one universe. In Chapter 11 I shall do the same for the common-sense idea that time 'flows', or that our consciousness 'moves' through time. In Chapter 3 I criticized inductivism, the common-sense idea that we form theories about the physical world by generalizing the results of observations, and that we justify our theories by repeating those observations. I explained that inductive generalization from observations is impossible, and that inductive justification is invalid. I explained that inductivism rests upon a mistaken idea of science as seeking predictions on the basis of observations, rather than as seeking explanations in response to problems. I also explained (following Popper) how science does make progress, by conjecturing new explanations and then choosing between the best ones by experiment. All this is largely accepted by scientists and philosophers of science. What is not accepted by most philosophers is that this process is *justified*. Let me explain.

Science seeks better explanations. A scientific explanation accounts for our observations by postulating something about what reality is like and how it works. We deem an explanation to be better if it leaves fewer loose ends (such as entities whose properties are themselves unexplained), requires fewer and simpler postulates, is more general, meshes more easily with good explanations in other fields and so on. But why should a better *explanation* be what we always assume it to be in practice, namely the token of a *truer theory*? Why, for that matter, should a downright bad explanation (one that has none of the above attributes, say) necessarily be false? There is indeed no logically necessary connection between truth and explanatory power. A bad explanation (such as solipsism) *may* be true. Even the best and truest available theory may make a false prediction in particular cases, and those might be the very cases in which we rely on the theory. No valid form of reasoning can logically rule out such possibilities, or even prove them unlikely. But in that case, what justifies our relying on our best explanations as guides to practical decision-making? More generally, *whatever* criteria we used to judge scientific theories, how could the fact that a theory satisfied those criteria today possibly imply anything about what will happen if we rely on the theory tomorrow?

This is the modern form of the 'problem of induction'. Most philosophers are now content with Popper's contention that new theories are not inferred from anything, but are merely hypotheses. They also accept that scientific progress is made through conjectures and refutations (as described in Chapter 3), and that theories are accepted when all their rivals are refuted, and not by virtue of numerous confirming instances. They accept that the knowledge obtained in this way tends, in the event, to be reliable. The problem is that they do not see why it should be. Traditional inductivists tried to formulate a 'principle of induction', which said that confirming instances made a theory more likely, or that 'the future will resemble the past', or some such statement. They also tried to formulate an inductive scientific methodology, laying down rules for what sort of inferences one could validly draw from 'data'. They all failed, for the reasons I have explained. But even if they had succeeded, in the sense of constructing a scheme that could be followed successfully to create scientific knowledge, this would not have solved the problem of induction as it is nowadays understood. For in that case 'induction' would simply be another possible way of choosing theories, and the problem would remain of *why those theories should be a reliable basis for action*. In other words, philosophers who worry about this 'problem of induction' are not inductivists in the old-fashioned sense. They do not try to obtain or justify any theories inductively. They do not expect the sky to fall in, but they do not know how to justify that expectation.

Philosophers today yearn for this missing justification. They no longer believe that induction would provide it, yet they have an induction-shaped gap in their scheme of things, just as religious people who have lost their faith suffer from a 'God-shaped gap' in *their* scheme of things. But in my opinion there is little difference between having an *X*-shaped gap in one's scheme of things and believing in *X*. Hence to fit in with the more sophisticated conception of the problem of induction, I wish to redefine the term 'inductivist' to mean someone who believes that the *invalidity* of inductive justification is a problem for the foundations of science. In other words, an inductivist believes that there is a gap which must be filled, if not by a principle of induction then by something else. Some inductivists do not mind being so designated. Others do, so I shall call them *crypto-inductivists*.

Most contemporary philosophers are crypto-inductivists. What makes matters worse is that (like many scientists) they grossly underrate the role of explanation in the scientific process. So do most Popperian anti-inductivists, who are thereby led to deny that there is any such thing as justification (even tentative justification). This opens up a new explanatory gap in *their* scheme of things. The philosopher John Worrall has dramatized the problem as he sees it in an imaginary dialogue between Popper and several other philosophers, entitled 'Why Both Popper and Watkins Fail to Solve the Problem of Induction'.* The setting is the top of the Eiffel Tower. One of the participants – the 'Floater' – decides to descend by jumping over the side instead of using the lift in the usual way. The others try to persuade the Floater that jumping off means certain death. They use the best available scientific and philosophical arguments. But the infuriating Floater still expects to float down safely, and keeps pointing out that no rival expectation can logically be proved to be preferable on the basis of past experience.

I believe that we can justify our expectation that the Floater would be killed. The justification (always tentative, of course) comes from the explanations provided by the relevant scientific theories. To the extent that those explanations are good, it is rationally justified to rely on the predictions of corresponding theories. So, in reply to Worrall, I now present a dialogue of my own, set in the same place.

DAVID: Since I read what Popper has to say about induction, I have believed that he did indeed, as he claimed, solve the problem of induction. But few philosophers agree. Why?

CRYPTO-INDUCTIVIST: Because Popper never addressed the problem of induction as we understand it. What he did was present a critique of *inductivism*. Inductivism said that there is an 'inductive' form of reasoning which can derive, and justify the use of, general theories about the future, given evidence in the form of individual observations made in the past. It held that there was a principle of nature, the *principle of induction*, which said something like 'observations made in the future are likely to resemble observations made under similar circumstances in the past'. Attempts were made to formulate this in such a way that it would indeed allow one to derive, or justify, general theories from individual observations. They all failed. Popper's critique, though influential among scientists (especially in conjunction with his other work, elucidating the methodology of science), was hardly original. The unsoundness of inductivism had been

* In *Freedom and Rationality: Essays in Honour of John Watkins.*

known almost since it was invented, and certainly since David Hume's critique of it in the early eighteenth century. The problem of induction is not how to justify or refute the principle of induction, but rather, taking for granted that it is invalid, *how to justify any conclusion about the future from past evidence.* And before you say that one doesn't need to . . .

DAVID: One doesn't need to.

CRYPTO-INDUCTIVIST: But one does. This is what is so irritating about you Popperians: you deny the obvious. Obviously the reason why you are not even now leaping over this railing is, in part, that you consider it *justified* to rely on our best theory of gravity and *unjustified* to rely on certain other theories. (Of course, by 'our best theory of gravity' in this case I mean more than just general relativity. I am also referring to a complex set of theories about such things as air resistance, human physiology, the elasticity of concrete and the availability of mid-air rescue devices.)

DAVID: Yes, I would consider it justified to rely on that theory. According to Popperian methodology, one should in these cases rely on the *best-corroborated* theory – that is, the one that has been subjected to the most stringent tests and has survived them while its rivals have been refuted.

CRYPTO-INDUCTIVIST: You say 'one *should*' rely on the best-corroborated theory, but why, exactly? Presumably because, according to Popper, the process of corroboration has justified the theory, in the sense that its predictions are more likely to be true than the predictions of other theories.

DAVID: Well, not more likely than *all* other theories, because no doubt one day we'll have even better theories of gravity . . .

CRYPTO-INDUCTIVIST: Now look. Please let's agree not to trip each other up with quibbles that do not bear on the substance of what we are discussing. *Of course* there may be a better theory of gravity one day, but you have to decide whether to jump now, *now*. And given the evidence available to you now, you have chosen a certain theory to act upon. And you have chosen it according to Popperian criteria because you believe that those criteria are the ones most likely to select theories which make true predictions.

DAVID: Yes.

CRYPTO-INDUCTIVIST: So to summarize, you believe that the evidence currently available to you justifies the prediction that you would be killed if you leapt over the railing.

DAVID: No, it doesn't.

CRYPTO-INDUCTIVIST: But dammit, you are contradicting yourself. Just now you said that that prediction *is* justified.

DAVID: It is justified. But it was not justified by the evidence, if by 'the evidence' you mean all the experiments whose outcomes the theory correctly predicted in the past. As we all know, that evidence is consistent with an infinity of theories, including theories predicting every logically possible outcome of my jumping over the railing.

CRYPTO-INDUCTIVIST: So in view of that, I repeat, the whole problem is to find what does justify the prediction. That is the problem of induction.

DAVID: Well, that is the problem that Popper solved.

CRYPTO-INDUCTIVIST: That's news to me, and I've studied Popper extensively. But anyway, what is the solution? I'm eager to hear it. What justifies the prediction, if it isn't the evidence?

DAVID: Argument.

CRYPTO-INDUCTIVIST: Argument?

DAVID: Only argument ever justifies anything – tentatively, of course. All theorizing is subject to error, and all that. But still, argument can sometimes justify theories. That is what argument is for.

CRYPTO-INDUCTIVIST: I think this is another of your quibbles. You can't mean that the theory was justified by *pure* argument, like a mathematical theorem.* The evidence played some role, surely.

DAVID: Of course. This is an empirical theory, so, according to

* Actually mathematical theorems are not proved by 'pure' argument (independent of physics) either, as I shall explain in Chapter 10.

Popperian scientific methodology, crucial experiments play a pivotal role in deciding between it and its rivals. The rivals were refuted; it survived.

CRYPTO-INDUCTIVIST: And in consequence of that refuting and surviving, all of which happened in the past, the practical use of the theory to predict the future is now justified.

DAVID: I suppose so, though it seems misleading to say 'in consequence of' when we are not talking about a logical deduction.

CRYPTO-INDUCTIVIST: Well that's the whole point again: *what sort of consequence was it?* Let me try to pin you down here. You admit that it was both argument *and* the outcomes of experiments that justified the theory. If the experiments had gone differently, the argument would have justified a different theory. So do you accept that in that sense – yes, via the argument, but I don't want to keep repeating that proviso – the outcomes of past experiments did justify the prediction?

DAVID: Yes.

CRYPTO-INDUCTIVIST: So what exactly was it about those actual past outcomes that justified the prediction, as opposed to other possible past outcomes which might well have justified the contrary prediction?

DAVID: It was that the actual outcomes refuted all the rival theories, and corroborated the theory that now prevails.

CRYPTO-INDUCTIVIST: Good. Now listen carefully, because you have just said something which is not only provably untrue, but which you yourself conceded was untrue only moments ago. You say that the outcomes of experiments 'refuted all the rival theories'. But you know very well that no set of outcomes of experiments can refute all possible rivals to a general theory. You said yourself that any set of past outcomes is (I quote) 'consistent with an infinity of theories, including theories predicting every logically possible outcome of my jumping over the railing'. It follows inexorably that the prediction you favour *was not* justified by the experimental outcomes, because there are infinitely many other rivals to your theory, also unrefuted as yet, which make the opposite prediction.

DAVID: I'm glad I listened carefully, as you asked, for now I see that at least part of the difference between us has been caused by a misunderstanding over terminology. When Popper speaks of 'rival theories' to a given theory, he does not mean the set of all logically possible rivals: he means only the actual rivals, those proposed in the course of a rational controversy. (That includes theories 'proposed' purely mentally, by one person, in the course of a 'controversy' within one mind.)

CRYPTO-INDUCTIVIST: I see. Well, I'll accept your terminology. But incidentally (I don't think it matters, for present purposes, but I'm curious), isn't it a strange assertion you are attributing to Popper, that the reliability of a theory depends on the accident of what *other* theories – false theories – people have proposed in the past, rather than just on the content of the theory in question, and on the experimental evidence?

DAVID: Not really. Even you inductivists speak of . . .

CRYPTO-INDUCTIVIST: I am *not* an inductivist!

DAVID: Yes you are.

CRYPTO-INDUCTIVIST: Hmph! Once again, I shall accept your terminology if you insist. But you may as well call me a porcupine. It really is perverse to call a person an 'inductivist' if that person's whole thesis is that the *invalidity* of inductive reasoning presents us with an unsolved philosophical problem.

DAVID: I don't think so. I think that that thesis is what defines, and always has defined, an inductivist. But I see that Popper has at least achieved one thing: 'inductivist' has become a term of abuse! Anyway, I was explaining why it's not so strange that the reliability of a theory should depend on what false theories people have proposed in the past. Even inductivists speak of a theory being reliable or not, given certain 'evidence'. Well, Popperians might speak of a theory being the best available for use in practice, given a certain *problem-situation*. And the most important features of a problem-situation are: what theories and explanations are in contention, what arguments have been advanced, and what theories have been refuted. 'Corroboration' is not just the confirmation of the winning theory. It requires the experimental refutation of rival theories. Confirming instances in themselves have no significance.

CRYPTO-INDUCTIVIST: Very interesting. I now understand the role of a theory's refuted rivals in the justification of its predictions. Under inductivism, observation was supposed to be primary. One imagined a mass of past observations from which the theory was supposed to be induced, and observations also constituted the evidence which somehow justified the theory. In the Popperian picture of scientific progress, it is not observations but problems, controversies, theories and criticism that are primary. Experiments are designed and performed only to resolve controversies. Therefore only experimental results that actually do refute a theory – and not just any theory, it must have been a genuine contender in a rational controversy – constitute 'corroboration'. And so it is only those experiments that provide evidence for the reliability of the winning theory.

DAVID: Correct. And even then, the 'reliability' that corroboration confers is not absolute but only relative to the other contending theories. That is, we expect the strategy of relying on corroborated theories to select the best theories from those that are proposed. That is a sufficient basis for action. We do not need (and could not validly get) any assurance about *how good* even the best proposed course of action will be. Furthermore, we may always be mistaken, but so what? We cannot use theories that have yet to be proposed; nor can we correct errors that we cannot yet see.

CRYPTO-INDUCTIVIST: Quite so. I am glad to have learned something about scientific methodology. But now – and I hope you don't think me impolite – I must draw your attention yet again to the question I have been asking all along. Suppose that a theory has passed through this whole process. Once upon a time it had rivals. Then experiments were performed and all the rivals were refuted. But it itself was not refuted. Thus it was corroborated. *What is it about its being corroborated that justifies our relying on it in the future?*

DAVID: Since all its rivals have been refuted, they are no longer rationally tenable. The corroborated theory is the only rationally tenable theory remaining.

CRYPTO-INDUCTIVIST: But that only shifts the focus from the future import of past corroboration to the future import of past refutation. The same problem remains. Why, exactly, is an experimentally refuted theory 'not rationally tenable'? Is it that having even one false consequence implies that it cannot be true?

DAVID: Yes.

CRYPTO-INDUCTIVIST: But surely, as regards the future applicability of the theory, that is not a logically relevant criticism. Admittedly, a refuted theory cannot be true *universally** – in particular, it cannot have been true in the past, when it was tested. But it could still have many true consequences, and in particular it could be universally true in the future.

DAVID: This 'true in the past' and 'true in the future' terminology is misleading. Each specific prediction of a theory is either true or false; that cannot change. What you really mean is that though the refuted theory is strictly false, because it makes some false predictions, all its predictions about the future might nevertheless be true. In other words, a *different theory*, which makes the same predictions about the future but different predictions about the past, might be true.

CRYPTO-INDUCTIVIST: If you like. So instead of asking why a refuted theory is not rationally tenable, I should, strictly speaking, have asked this: why does the refutation of a theory also render untenable every variant of the theory that agrees with it about the future – even a variant that has not been refuted?

DAVID: It is not that refutation *renders* such theories untenable. It is just that sometimes they already *are* untenable, by virtue of being bad explanations. And that is when science can make progress. For a theory to win an argument, all its rivals must be untenable, and that includes all the variants of the rivals which anyone has thought of. But remember, it is only the rivals *which*

* Actually it could still be true universally, if other theories about the experimental set-up were false.

anyone has thought of that need be untenable. For example, in the case of gravity no one has ever proposed a tenable theory that agrees with the prevailing one in all its tested predictions, but differs in its predictions about future experiments. I am sure that such theories are possible – for instance, the successor to the prevailing theory will presumably be one of them. But if no one has yet thought of such a theory, how can anyone act upon it?

CRYPTO-INDUCTIVIST: What do you mean, 'no one has yet thought of such a theory'? I could easily think of one right now.

DAVID: I very much doubt that you can.

CRYPTO-INDUCTIVIST: Of course I can. Here it is. 'Whenever you, David, jump from high places in ways that would, according to the prevailing theory, kill you, you float instead. Apart from that, the prevailing theory holds universally.' I put it to you that every past test of your theory was also necessarily a test of mine, since all the predictions of your theory and mine regarding past experiments are identical. Therefore your theory's refuted rivals were also my theory's refuted rivals. And therefore my new theory is exactly as corroborated as your prevailing theory. How, then, can my theory be 'untenable'? What faults could it possibly have that are not shared by your theory?

DAVID: Just about every fault in the Popperian book! Your theory is constructed from the prevailing one by appending an unexplained qualification about me floating. That qualification is, in effect, a new theory, but you have given no argument either against the prevailing theory of my gravitational properties, or in favour of the new one. You have subjected your new theory to no criticism (other than what I am giving it now) and no experimental testing. It does not solve – or even purport to solve – any current problem, nor have you suggested a new, interesting problem that it could solve. Worst of all, your qualification explains nothing, but *spoils the explanation* of gravity that is the basis of the prevailing theory. It is this explanation that justifies our relying on the prevailing theory and not on yours. Thus by all rational criteria your proposed qualification can be summarily rejected.

CRYPTO-INDUCTIVIST: Couldn't I say exactly the same thing about your theory? Your theory differs from mine only by the same minor qualification, but in reverse. You think I ought to have explained my qualification. But why are our positions not symmetrical?

DAVID: Because your theory does not come with an explanation of its predictions, and mine does.

CRYPTO-INDUCTIVIST: But if my theory had been proposed first, it would have been your theory that appeared to contain an unexplained qualification, and it would be your theory that would be 'summarily rejected'.

DAVID: That is simply untrue. Any rational person who was comparing your theory with the prevailing one, even if yours had been proposed first, would immediately reject your theory in favour of the prevailing one. For the fact that your theory is an unexplained modification of another theory is manifest in your very statement of it.

CRYPTO-INDUCTIVIST: You mean that my theory has the form 'such-and-such a theory holds universally, except in such-and-such a situation', but I don't explain why the exception holds?

DAVID: Exactly.

CRYPTO-INDUCTIVIST: Aha! Well, I think I can prove you wrong here (with the help of the philosopher Nelson Goodman). Consider a variant of the English language that has no verb 'to fall'. Instead it has a verb 'to x-fall' which means 'to fall' except when applied to you, in which case it means 'to float'. Similarly, 'to x-float' means 'to float' except when applied to you, in which case it means 'to fall'. In this new language I could express my theory as the unqualified assertion 'all objects x-fall if unsupported'. But the prevailing theory (which in English says 'all objects fall if unsupported') would, in the new language, have to be qualified: 'all objects x-fall when unsupported, *except David, who x-floats*'. So which of these two theories is qualified depends on the language they are expressed in, doesn't it?

DAVID: In form, yes. But that is a triviality. Your theory contains, *in substance*, an unexplained assertion, qualifying the prevailing theory. The prevailing theory is, *in substance*, your theory stripped of an unexplained qualification. No matter how you slice it, that is an objective fact, independent of language.

CRYPTO-INDUCTIVIST: I don't see why. You yourself used the *form* of my theory to spot the 'unnecessary qualification'. You said that it was 'manifest' as an additional clause in my very statement of the theory – in English. But when the theory is translated into my language, no qualification is manifest; and on the contrary, a manifest qualification appears in the very statement of the prevailing theory.

DAVID: So it does. But not all languages are equal. *Languages are theories*. In their vocabulary and grammar, they embody substantial assertions about the world. Whenever we state a theory, only a small part of its content is explicit: the rest is carried by the language. Like all theories, languages are invented and selected for their ability to solve certain problems. In this case the problems are those of expressing other theories in forms in which it is convenient to apply them, and to compare and criticize them. One of the most important ways in which languages solve these problems is to embody, implicitly, theories that are uncontroversial and taken for granted, while allowing things that need to be stated or argued about to be expressed succinctly and cleanly.

CRYPTO-INDUCTIVIST: I accept that.

DAVID: Thus it is no accident when a language chooses to cover the conceptual ground with one set of concepts rather than another. It reflects the current state of the speakers' problem-situation. That is why the form of your theory, *in English*, is a good indication of its status *vis à vis* the current problem-situation – whether it solves problems or exacerbates them. But it is not the form of your theory I am complaining about. It is the substance. My complaint is that your theory solves nothing and only exacerbates the problem-situation. This defect is manifest when the theory is expressed in English, and implicit when it is expressed in your language. But it is no less severe for that. I could state my complaint equally well in English, or in scientific jargon, or in your proposed language or in any language capable of expressing the discussion we have been having. (It is a Popperian maxim that one should always be willing to carry on the discussion in the opponent's terminology.)

CRYPTO-INDUCTIVIST: You may have a point there. But could you elaborate? In what way does my theory exacerbate the problem-situation, and why would this be obvious even to a native speaker of my hypothetical language?

DAVID: Your theory asserts the existence of a physical *anomaly* which is not present according to the prevailing theory. The anomaly is my alleged immunity from gravity. Certainly, you can invent a language which expresses this anomaly implicitly, so that statements of your theory of gravity need not refer to it explicitly. But refer to it they do. A rose by any other name would smell as sweet. Suppose that you – indeed suppose that everyone – were a native speaker of your language, and believed your theory of gravity to be true. Suppose that we all took it entirely for granted, and thought it so natural that we used the same word 'x-fall' to describe what you or I would do if we jumped over the railing. None of that alters in the slightest degree the obvious difference there would be between my response to gravity and everything else's. If you fell over the railing, you might well envy me on the way down. You might well think, 'if only I could respond to gravity as David does, rather than in this entirely different way that I do!'

CRYPTO-INDUCTIVIST: That's true. Just because the same word 'x-falling' describes your response to gravity and mine, I wouldn't think that the actual response is the same. On the contrary, being a fluent speaker of this supposed language, I'd know very well that 'x-falling' was physically different for you and for me, just as a native English speaker knows that the words 'being drunk' mean something physically different for a person and for a glass of water. I wouldn't think, 'if this had happened to David, he'd be x-falling just as I am'. I'd think, 'if this had happened to David, he'd x-fall and survive, while I shall x-fall and die.'

DAVID: Moreover, despite your being sure that I would float, *you wouldn't understand why*. Knowing is not the same as understanding. You would be curious as to the explanation of this 'well-known' anomaly. So would everyone else. Physicists would congregate from all over the world to study my anomalous gravitational properties. In fact, if your language were really the prevailing one, and your theory were really taken for granted by everyone, the scientific world would presumably have been impatiently awaiting my very birth, and would be queuing for the privilege of dropping me out of aircraft! But of course, the premise of all this, namely that your theory is taken for granted and embodied in the prevailing language, is preposterous. Theory or no theory, language or no language, in reality no rational person would entertain the possibility of such a glaring physical anomaly without there being a very powerful explanation in its favour. Therefore, just as your theory would be summarily rejected, your language would be rejected too, for it is just another way of stating your theory.

CRYPTO-INDUCTIVIST: Could it be that there is a solution of the problem of induction lurking here after all? Let me see. How does this insight about language change things? My argument relied upon an apparent symmetry between your position and mine. We both adopted theories that were consistent with existing experimental results, and whose rivals (except each other) had been refuted. You said that I was being irrational because my theory involved an unexplained assertion, but I countered by saying that in a different language it would be your theory that contained such an assertion, so the symmetry was still there. But now you have pointed out that languages are theories, and that the combination of my proposed language and theory assert the existence of an objective, physical anomaly, as compared with what the combination of the English language and the prevailing theory assert. This is where the symmetry between our positions, and the argument I was putting forward, break down hopelessly.

DAVID: Indeed they do.

CRYPTO-INDUCTIVIST: Let me see if I can clarify this a little further. Are you saying that it is a principle of rationality that a theory which asserts the existence of an objective, physical anomaly is, other things being equal, less likely to make true predictions than one that doesn't?

DAVID: Not quite. Theories postulating anomalies *without explaining them* are less likely *than their rivals* to make true predictions. More generally, it is a principle of rationality that theories are postulated in order to solve problems. Therefore *any* postulate which solves no problem is to be rejected. That is because a good explanation qualified by such a postulate becomes a bad explanation.

CRYPTO-INDUCTIVIST: Now that I understand that there really is an objective difference between theories which make unexplained predictions and theories which don't, I must admit that this does look promising as a solution of the problem of induction. You seem to have discovered a way of justifying your future reliance on the theory of gravity, given only the past problem-situation (including past observational evidence) and the distinction between a good explanation and a bad one. You do not have to make any assumption such as 'the future is likely to resemble the past'.

DAVID: It was not I who discovered this.

CRYPTO-INDUCTIVIST: Well, I don't think Popper did either. For one thing, Popper did not think that scientific theories could be *justified* at all. You make a careful distinction between theories being justified by observations (as inductivists think) and being justified by argument. But Popper made no such distinction. And in regard to the problem of induction, he actually said that although future predictions of a theory cannot be justified, we should act as though they were!

DAVID: I don't think he said that, exactly. If he did, he didn't really mean it.

CRYPTO-INDUCTIVIST: *What?*

DAVID: Or if he did mean it, he was mistaken. Why are you so upset? It is perfectly possible for a person to discover a new theory (in this case Popperian epistemology) but nevertheless to continue to hold beliefs that contradict it. The more profound the theory is, the more likely this is to happen.

CRYPTO-INDUCTIVIST: Are you claiming to understand Popper's theory better than he did himself?

DAVID: I neither know nor care. The reverence that philosophers show for the historical sources of ideas is very perverse, you know. In science we do not consider the discoverer of a theory to have any special insight into it. On the contrary, we hardly ever consult original sources. They invariably become obsolete, as the problem-situations that prompted them are transformed by the discoveries themselves. For example, most relativity theorists today understand Einstein's theory better than he did. The founders of quantum theory made a complete mess of understanding their own theory. Such shaky beginnings are to be expected; and when we stand upon the shoulders of giants, it may not be all that hard to see further than they did. But in any case, surely it is more interesting to argue about what the truth is, than about what some particular thinker, however great, did or did not think.

CRYPTO-INDUCTIVIST: All right, I agree. But wait a moment, I think I spoke too soon when I said that you were not postulating any sort of principle of induction. Look: you have justified a theory about the future (the prevailing theory of gravity) as being more reliable than another theory (the one I proposed), even though they are both consistent with all currently known observations. Since the prevailing theory applies both to the future and to the past, you have justified the proposition that, as regards gravity, *the future resembles the past*. And the same would hold whenever you justify a theory as reliable on the grounds that it is corroborated. Now, in order to go from 'corroborated' to 'reliable', you examined the theories' explanatory power. So what you have shown is that what we might call the 'principle of seeking better explanations', together with some observations – yes, and arguments – *imply* that the future will, in many respects, resemble the past. And that is a principle of induction! If your 'explanation principle' implies a principle of induction, then, logically, it *is* a principle of induction. So inductivism is true after all, and a principle of induction does indeed have to be postulated, explicitly or implicitly, before we can predict the future.

DAVID: Oh dear! This inductivism really is a virulent disease. Having gone into remission for only a few seconds, it now returns more violently than before.

CRYPTO-INDUCTIVIST: Does Popperian rationalism justify *ad hominem* arguments as well? I ask for information only.

DAVID: I apologize. Let me go straight to the substance of what you said. Yes, I have justified an assertion about the future. You say this implies that 'the future resembles the past'. Well, vacuously, yes, inasmuch as *any* theory about the future would assert that it resembled the past in some sense. But this inference that the future resembles the past is not the sought-for principle of induction, for we could neither derive nor justify any theory or prediction about the future from it. For example, we could not use it to distinguish your theory of gravity from the prevailing one, for they both say, in their own way, that the future resembles the past.

CRYPTO-INDUCTIVIST: Couldn't we derive, from the 'explanation principle', a form of the principle of induction that *could* be used to select theories? What about: 'if an unexplained anomaly does not happen in the past, then it is unlikely in the future'?

DAVID: No. Our justification does not depend on whether a particular anomaly happens in the past. It has to do with whether there is an explanation for the existence of that anomaly.

CRYPTO-INDUCTIVIST: All right then, let me formulate it more carefully: 'if, in the present, there is no explanatory theory predicting that a particular anomaly will happen in the future, then that anomaly is unlikely to happen in the future'.

DAVID: That may well be true. I, for one, believe that it is. However, it is not of the form 'the future is likely to resemble the past'. Moreover, in trying to make it look as much like that as possible, you have specialized it to cases 'in the present', 'in the future', and to the case of an 'anomaly'. But it is just as true without these specializations. It is just a general statement about the efficacy of argument. In short, if there is no argument in favour of a postulate, then it is not reliable. Past, present or future. Anomaly or no anomaly. Period.

CRYPTO-INDUCTIVIST: Yes, I see.

DAVID: Nothing in the concepts of 'rational argument' or 'explanation' relates the future to the past in any special way. Nothing is postulated about anything 'resembling' anything. Nothing of that sort would help if it were postulated. In the vacuous sense in which the very concept of 'explanation' implies that the future 'resembles the past', it nevertheless implies nothing specific about the future, so it is not a principle of induction. There is no principle of induction. There is no process of induction. No one ever uses them or anything like them. And there is no longer a problem of induction. Is that clear now?

CRYPTO-INDUCTIVIST: Yes. Please excuse me for a few moments while I adjust my entire world-view.

DAVID: To assist you in that exercise, I think you should consider your alternative 'theory of gravity' more closely.

CRYPTO-INDUCTIVIST: . . .

DAVID: As we have agreed, your theory consists objectively of a theory of gravity (the prevailing theory), qualified by an unexplained prediction about me. It says that I would float, unsupported. 'Unsupported' means 'without any upward force acting' on me, so the suggestion is that I would be immune to the 'force' of gravity which would otherwise pull me down. But according to the general theory of relativity, gravity is not a force but a manifestation of the curvature of spacetime. This curvature explains why unsupported objects, like myself and the Earth, move closer together with time. Therefore, in the light of modern physics your theory is presumably saying that there *is* an upward force on me, as required to hold me at a constant distance from the Earth. But where does that force come from, and how does it behave? For example, what is a 'constant distance'? If the Earth were to move downwards, would I respond instantaneously to maintain the same height (which would allow communication faster than the speed of light, contrary to another principle of relativity), or would the information about where the Earth is have to reach me at the speed of light first? If so, what carries this information? Is it a new sort of wave emitted by the Earth – in which case what equations does it obey? Does it carry energy? What is its quantum-mechanical behaviour? Or is it that I respond in a special way to existing waves, such as light? In that case, would the anomaly disappear if an opaque barrier were placed between me and the Earth? Isn't the Earth mostly opaque anyway? Where does 'the Earth' begin: what defines the surface above which I am supposed to 'float'?

CRYPTO-INDUCTIVIST: ...

DAVID: For that matter, what defines where *I* begin? If I hold on to a heavy weight, does it float too? If so, then the aircraft in which I have flown could have switched off their engines without mishap. What counts as 'holding on'? Would the aircraft then drop if I let go of the arm rest? And if the effect does not apply to things I am holding on to, what about my clothes? Will they weigh me down and cause me to be killed after all, if I jump over the railing? What about my last meal?

CRYPTO-INDUCTIVIST: ...

DAVID: I could go on like this *ad infinitum*. The point is, the more we consider the implications of your proposed anomaly, the more unanswered questions we find. This is not just a matter of your theory being incomplete. These questions are *dilemmas*. Whichever way they are answered, they create fresh problems by spoiling satisfactory explanations of other phenomena.

CRYPTO-INDUCTIVIST: ...

DAVID: So your additional postulate is not just superfluous, it is positively bad. In general, perverse but unrefuted theories which one can propose off the cuff fall roughly into two categories. There are theories that postulate unobservable entities, such as particles that do not interact with any other matter. They can be rejected for solving nothing ('Occam's razor', if you like). And there are theories, like yours, that predict unexplained observable anomalies. They can be rejected for solving nothing *and* spoiling existing solutions. It is not, I hasten to add, that they conflict with existing observations. It is that they remove the explanatory power from existing theories by asserting that the predictions of those theories have exceptions, but not explaining how. You can't just say 'spacetime geometry brings unsupported objects together, *unless* one of them is David, in which case it leaves them alone'. Either the explanation of gravity is spacetime curvature or it isn't. Just compare your theory with the perfectly legitimate assertion that a feather would float down slowly because there would indeed be a sufficient upward force on it from the air. That assertion is a consequence of our existing explanatory theory of what air is, so it raises no new problem, as your theory does.

CRYPTO-INDUCTIVIST: I see that. Now, will you give me some help in adjusting my world-view?

DAVID: Well, have you read my book, *The Fabric of Reality*?

CRYPTO-INDUCTIVIST: I certainly plan to, but for the moment the help that I was asking for concerns a very specific difficulty.

DAVID: Go ahead.

CRYPTO-INDUCTIVIST: The difficulty is this. When I rehearse the discussion we have been having, I am entirely convinced that your prediction of what would happen if you or I jumped off this tower was not derived from any inductive hypothesis such as 'the future resembles the past'. But when I step back and consider the overall logic of the situation, I fear I still cannot understand how that can be. Consider the raw materials for the argument. Initially, I assumed that past observations and deductive logic are our only raw material. Then I admitted that the current problem-situation is relevant too, because we need justify our theory only as being more reliable than existing rivals. And then I had to take into account that vast classes of theories can be ruled out by argument alone, because they are bad explanations, and that the principles of rationality can be included in our raw material. What I cannot understand is where in that raw material – *past* observations, the *present* problem-situation and *timeless* principles of logic and rationality, none of which justifies inferences from the past to the future – the justification of future predictions has come from. There seems to be a logical gap. Are we making a hidden assumption somewhere?

DAVID: No, there is no logical gap. What you call our 'raw material' does indeed include assertions about the future. The best existing theories, which cannot be abandoned lightly because they are the solutions of problems, contain predictions about the future. And these predictions cannot be severed from the theories' other content, as you tried to do, because that would spoil the theories' explanatory power. Any new theory we propose must therefore *either* be consistent with these existing theories, which has implications for what the new theory can say about the future, *or* contradict some existing theories but address the problems thereby raised, giving alternative explanations, which again constrains what they can say about the future.

CRYPTO-INDUCTIVIST: So we have no principle of reasoning which says that the future will resemble the past, but we do have actual theories which say that. So do we have actual theories which imply a limited form of inductive principle?

DAVID: No. Our theories simply assert something about the future. Vacuously, any theory about the future implies that the future will 'resemble the past' in some ways. But we only find out in what respects the theory says that the future will resemble the past after we have the theory. You might as well say that since our theories hold certain features of reality to be the same throughout *space*, they imply a 'spatial principle of induction' to the effect that 'the near resembles the distant'. Let me point out that, in any practical sense of the word 'resemble', our present theories say that the future will *not* resemble the past. The cosmological 'Big Crunch', for instance (the recollapse of the universe to a single point), is an event that some cosmologists predict, but which is just about as unlike the present epoch, in every physical sense, as it could possibly be. The very laws from which we predict its occurrence will not apply to it.

CRYPTO-INDUCTIVIST: I am convinced on that point. Let me try one last argument. We have seen that future predictions can be justified by appeal to the principles of rationality. But what justifies those? They are not, after all, truths of pure logic. So there are two possibilities: either they are unjustified, in which case conclusions drawn from them are unjustified too; or they are justified by some as yet unknown means. In either case there is a missing justification. I no longer suspect that this is the problem of induction in disguise. Nevertheless, having exploded the problem of induction, have we not revealed another fundamental problem, also concerning missing justification, beneath?

DAVID: What justifies the principles of rationality? Argument, as usual. What, for instance, justifies our relying on the laws of *deduction*, despite the fact that any attempt to justify them logically must lead either to circularity or to an infinite regress? They are justified because no explanation is improved by replacing a law of deduction.

CRYPTO-INDUCTIVIST: That doesn't seem a very secure foundation for pure logic.

DAVID: It is not perfectly secure. Nor should we expect it to be, for logical reasoning is no less a physical process than scientific reasoning is, and it is inherently fallible. The laws of logic are not self-evident. There are people, the mathematical 'intuitionists', who disagree with the conventional laws of deduction (the logical 'rules of inference'). I discuss their strange world-view in Chapter 10 of *The Fabric of Reality*. They cannot be *proved* wrong, but I shall *argue* that they are wrong, and I am sure you will agree that my argument justifies this conclusion.

CRYPTO-INDUCTIVIST: So you don't think that there is a 'problem of deduction', then?

DAVID: No. I don't think that there is a problem with any of the usual ways of justifying conclusions in science, philosophy or mathematics. However, it is an interesting *fact* that the physical universe admits processes that create knowledge about itself, and about other things too. We may reasonably try to explain this fact in the same way as we explain other physical facts, namely through explanatory theories. You will see in Chapter 6 of *The Fabric of Reality* that I think that the Turing principle is the appropriate theory in this case. It says that it is possible to build a virtual-reality generator whose repertoire includes every physically possible environment. If the Turing principle is a law of physics, as I have argued that it is, then we should not be surprised to find that we can form accurate theories about reality, because that is just virtual reality in action. Just as the fact that steam engines are possible is a direct expression of the principles of thermodynamics, so the fact that the human brain is capable of creating knowledge is a direct expression of the Turing principle.

CRYPTO-INDUCTIVIST: But how do we know that the Turing principle is *true*?

DAVID: We don't, of course ... But you are afraid, aren't you, that if we can't justify the Turing principle, then we shall once again have lost our justification for relying on scientific predictions?

CRYPTO-INDUCTIVIST: Er, yes.

DAVID: But we have now moved on to a completely different question! We are now discussing an apparent *fact* about physical reality, namely that it can make reliable predictions about itself. We are trying to explain that fact, to place it within the same framework as other facts we know. I suggested that there may be a certain law of physics involved. But if I were wrong about that, indeed even if we were entirely unable to explain this remarkable property of reality, that would not detract one jot from the justification of any scientific theory. For it would not make the explanations in such a theory one jot worse.

CRYPTO-INDUCTIVIST: Now my arguments are exhausted. Intellectually, I am convinced. Yet I must confess that I still feel what I can only describe as an 'emotional doubt'.

DAVID: Perhaps it will help if I make one last comment, not about any of the specific arguments you have raised, but about a misconception that seems to underlie many of them. You know that it is a misconception; yet you may not yet have incorporated the ramifications of that into your world-view. Perhaps that is the source of your 'emotional doubt'.

CRYPTO-INDUCTIVIST: Fire away.

DAVID: The misconception is about the very nature of argument and explanation. You seem to be assuming that arguments and explanations, such as those that justify acting on a particular theory, have the form of mathematical proofs, proceeding from assumptions to conclusions. You look for the 'raw material' (axioms) from which our conclusions (theorems) are derived. Now, there is indeed a logical structure of this type associated with every successful argument or explanation. But the process of argument does not begin with the 'axioms' and end with the 'conclusion'. Rather, it starts in the middle, with a version that is riddled with inconsistencies, gaps, ambiguities and irrelevancies. All these faults are criticized. Attempts are made to replace faulty theories. The theories that are criticized and replaced usually include some of the 'axioms'. That is why it is a mistake to assume that an argument begins with, or is justified by, the theories that eventually serve as its 'axioms'. The argument ends – tentatively – when it seems to have shown that the associated explanation is satisfactory. The 'axioms' adopted are not ultimate, unchallengeable beliefs. They are tentative, explanatory theories.

CRYPTO-INDUCTIVIST: I see. Argument is not the same species of thing as deduction, or the non-existent induction. It is not based on anything or justified by anything. And it *doesn't have to be*, because its purpose is to solve problems – to show that a given problem is solved by a given explanation.

DAVID: Welcome to the club.

EX-INDUCTIVIST: All these years I have felt so secure in my great Problem. I felt so superior both to the ancient inductivists, and to the upstart Popper. And all the time, without even knowing it, I was a crypto-inductivist myself! Inductivism is indeed a disease. It makes one blind.

DAVID: Don't be too hard on yourself. You are cured now. If only your fellow-sufferers were as amenable to being cured by mere argument!

EX-INDUCTIVIST: But how could I have been so blind? To think that I once nominated Popper for the Derrida Prize for Ridiculous

Pronouncements, while all the time he had solved the problem of induction! O *mea culpa*! God save us, for we have burned a saint! I feel so ashamed. I see no way out but to throw myself over this railing.

DAVID: Surely that is not called for. We Popperians believe in letting our theories die in our place. Just throw *inductivism* overboard instead.

EX-INDUCTIVIST: I will, I will!

TERMINOLOGY

crypto-inductivist Someone who believes that the invalidity of inductive reasoning raises a serious philosophical problem, namely the problem of how to justify relying on scientific theories.

Next, the fourth strand, the theory of evolution, which answers the question 'what is life?'

8

The Significance of Life

From ancient times until about the nineteenth century, it was taken for granted that some special animating force or factor was required to make the matter in living organisms behave so noticeably differently from other matter. This would mean in effect that there were two types of matter in the universe: *animate* matter and *inanimate* matter, with fundamentally different physical properties. Consider a living organism such as a bear. A photograph of a bear resembles the living bear in some respects. So do other inanimate objects such as a dead bear, or even, in a very limited fashion, the Great Bear constellation. But only animate matter can chase you through the forest as you dodge round trees, and catch you and tear you apart. Inanimate things never do anything as purposeful as that – or so the ancients thought. They had, of course, never seen a guided missile.

To Aristotle and other ancient philosophers, the most conspicuous feature of animate matter was its ability to initiate motion. They thought that when inanimate matter, such as a rock, has come to rest, it never moves again unless something kicks it. But animate matter, such as a hibernating bear, can be at rest and then begin to move without being kicked. With the benefit of modern science we can easily pick holes in these generalizations, and the very idea of 'initiating motion' now seems misconceived: we know that the bear wakes up because of electrochemical processes in its body. These may be initiated by external 'kicks' such as rising temperature, or by an internal biological clock which uses slow chemical reactions to keep time. Chemical reactions are nothing more than the motion of atoms, so the bear never is entirely at rest. On the other hand a uranium nucleus, which is certainly not alive, may remain unchanged for billions of years and then, without any stimulus at all, suddenly and violently disintegrate. So the nominal content of Aristotle's idea is worthless today. But he did get one important thing right which most modern thinkers have got wrong. In trying to associate life with a basic physical concept (albeit the wrong one, motion), he recognized that life is a fundamental phenomenon of nature.

A phenomenon is 'fundamental' if a sufficiently deep understanding of the world depends on understanding that phenomenon. Opinions differ, of course, about what aspects of the world are worth understanding, and consequently about what is deep or fundamental. Some would say that love is the most fundamental phenomenon in the world. Others believe that when one has learned certain sacred texts by heart, one understands everything that is worth understanding. The understanding that I am talking about is expressed in laws of physics, and in principles of logic and philosophy. A 'deeper' understanding is one that has more generality, incorporates more connections between superficially diverse truths, explains more with fewer unexplained assumptions. The most fundamental phenomena are implicated in the explanation of many other phenomena, but are themselves explained only by basic laws and principles.

Not all fundamental phenomena have large physical effects. Gravitation does, and is indeed a fundamental phenomenon. But the direct effects of quantum interference, such as the shadow patterns described in Chapter 2, are not large. It is quite hard even to detect them unambiguously. Nevertheless, we have seen that quantum interference is a fundamental phenomenon. Only by understanding it can we understand the basic fact about physical reality, namely the existence of parallel universes.

It was obvious to Aristotle that life is theoretically fundamental *and* has large physical effects. As we shall see, he was right. But it was obvious to him for quite the wrong reasons, namely the supposedly distinctive mechanical properties of animate matter, and the domination of the Earth's surface by living processes. Aristotle thought that the universe consists principally of what we now call the biosphere (life-containing region) of the Earth, with a few extra bits – celestial spheres and the Earth's interior – tacked on above and below. If the Earth's biosphere is the principal component of your cosmos, you will naturally think that trees and animals are at least as important as rocks and stars in the great scheme of things, especially if you know very little physics or biology. Modern science has led to almost the opposite conclusion. The Copernican revolution made the Earth subsidiary to a central, inanimate Sun. Subsequent discoveries in physics and astronomy showed not only that the universe is vast in comparison with the Earth, but that it is described with enormous accuracy by all-encompassing laws that make no mention of life at all. Charles Darwin's theory of evolution explained the origin of life in terms that required no special physics, and since then we have discovered many of the detailed mechanisms of life, and found no special physics there either.

These spectacular successes of science, and the great generality of Newtonian and subsequent physics in particular, did much to make reductionism attractive. Since faith in revealed truth had been found to be incompatible with rationality (which requires an openness to criticism), many people nevertheless yearned for an ultimate foundation to things in which they could believe. If they did not yet have a reductive 'theory of everything' to believe in, then at least they aspired to one. It was taken for granted that a reductionist hierarchy of sciences, based on subatomic physics, was integral to the scientific world-view, and so it was criticized only by pseudo-scientists and others who rebelled against science itself. Thus, by the time I learned biology in school, the status of that subject had changed to the opposite of what Aristotle thought was obvious. Life was not considered to be fundamental at all. The very term 'nature study' – meaning biology – had become an anachronism. Fundamentally, nature was physics. I am oversimplifying only a little if I characterize the prevailing view as follows. Physics had an offshoot, chemistry, which studied the interactions of atoms. Chemistry had an offshoot, organic chemistry, which studied the properties of compounds of the element carbon. Organic chemistry in turn had an offshoot, biology, which studied the chemical processes we call life. Only because we happen to *be* such a process was this remote offshoot of a fundamental subject interesting to us. Physics, in contrast, was regarded as self-evidently important in its own right because the entire universe, life included, conforms to its principles.

My classmates and I had to learn by heart a number of 'characteristics of living things'. These were merely descriptive. They made little reference to fundamental concepts. Admittedly, (loco)*motion* was one of them – an ill-defined echo of the Aristotelian idea – but *respiration* and *excretion* were among them as well. There was also *reproduction*, *growth*, and the memorably named *irritability*, which meant that if you kick it, it kicks back. What these supposed characteristics of life lack in elegance and profundity, they do not make up in accuracy. As Dr Johnson would tell us, every real object is 'irritable'. On the other hand, viruses do not respire, grow, excrete, or move (unless kicked), but they are alive. And sterile human beings do not reproduce, yet they are alive too.

The reason why both Aristotle's view and that of my school textbooks failed to capture even a good taxonomic distinction between living and non-living things, let alone anything deeper, is that they both miss the point about what living things are (a mistake more forgivable in Aristotle because in his day no one knew any better). Modern biology does not try to define life by some characteristic physical attribute or substance – some living 'essence' – with which only animate matter is endowed. We no longer expect there to be any such essence, because we now know that 'animate matter', matter in the form of living organisms, is not the basis of life. It is merely one of the effects of life, and the basis of life is molecular. It is the fact that there exist molecules which cause certain environments to make copies of those molecules.

Such molecules are called *replicators*. More generally, a replicator is any entity that causes certain environments to copy it. Not all replicators are biological, and not all replicators are molecules. For example, a self-copying computer program (such as a computer virus) is a replicator. A good joke is another replicator, for it causes its listeners to retell it to further listeners. Richard Dawkins has coined the term *meme* (rhyming with 'cream') for replicators that are human ideas, such as jokes. But all life on Earth is based on replicators that are molecules. These are called *genes*, and biology is the study of the origin, structure and operation of genes, and of their effects on other matter. In most organisms a gene consists of a sequence of smaller molecules, of which there are four different kinds, joined together in a chain. The names of the component molecules (adenine, cytosine, guanine and thymine) are usually shortened to A, C, G and T. The abbreviated chemical name for a chain of any number of A, C, G and T molecules, in any order, is DNA.

Genes are in effect computer programs, expressed as sequences of A, C, G and T symbols in a standard language called the *genetic code* which, with very slight variations, is common to all life on Earth. (Some viruses are based on a related type of molecule, RNA, while prions are, in a sense, self-replicating protein molecules.) Special structures within each organism's cells act as computers to execute these gene programs. The execution consists of manufacturing certain molecules (proteins) from simpler molecules (amino acids) under certain external conditions. For example, the sequence 'ATG' is an instruction to incorporate the amino acid methionine into the protein molecule being manufactured.

Typically, a gene is chemically 'switched on' in certain cells of the body, and then instructs those cells to manufacture the corresponding protein. For example, the hormone insulin, which controls blood sugar levels in vertebrates, is such a protein. The gene for manufacturing it is present in almost every cell of the body, but it is switched on only in certain specialized cells in the pancreas, and then only when it is needed. At the molecular level, this is all that any gene can program its cellular computer to do: manufacture a certain chemical. But genes succeed in being replicators because these low-level chemical programs add up, through layer upon layer of complex control and feedback, to sophisticated high-level instructions. Jointly, the insulin gene and the genes involved in switching it on and off amount to a complete program for the regulation of sugar in the bloodstream.

Similarly, there are genes which contain specific instructions for how and when they and other genes are to be copied, and instructions for the manufacture of further organisms of the same species, including the molecular computers which will execute all these instructions again in the next generation. There are also instructions for how the organism as a whole should respond to stimuli – for instance, when and how it should hunt, eat, mate, fight or run away. And so on.

A gene can function as a replicator only in certain environments. By analogy with an ecological 'niche' (the set of environments in which an organism can survive and reproduce), I shall also use the term *niche* for the set of all possible environments which a given replicator would cause to make copies of it. The niche of an insulin gene includes environments where the gene is located in the nucleus of a cell in the company of certain other genes, and the cell itself is appropriately located within a functioning organism, in a habitat suitable for sustaining the organism's life and reproduction. But there are also other environments – such as biotechnology laboratories in which bacteria are genetically altered so as to incorporate the gene – which likewise copy the insulin gene. Those environments are also part of the gene's niche, as are an infinity of other possible environments that are very different from those in which the gene evolved.

Not everything that can be copied is a replicator. A replicator *causes* its environment to copy it: that is, it contributes causally to its own copying. (My terminology differs slightly from that used by Dawkins. Anything that is copied, for whatever reason, he calls a replicator. What I call a replicator he would call an *active* replicator.) What it means in general to contribute causally to something is an issue to which I shall return, but what I mean here is that the presence and specific physical form of the replicator *makes a difference* to whether copying takes place or not. In other words, the replicator is copied if it is present, but if it were replaced by almost any other object, even a rather similar one, that object would not be copied. For example, the insulin gene causes only one small step in the enormously complicated process of its own replication (that process being the whole life cycle of the organism). But the overwhelming majority of variants of that gene would not instruct cells to manufacture a chemical that could do the job of insulin. If the insulin genes in an individual organism's cells were replaced by slightly different molecules, that organism would die (unless it were kept alive by other means), and would therefore fail to have offspring, and those molecules would not be copied. So whether copying takes place or not is exquisitely sensitive to the physical form of the insulin gene. The presence of the gene in its proper form and location *makes a difference* to whether copying takes place, which makes it a replicator, though there are countless other causes contributing to its replication as well.

Along with genes, *random* sequences of A, C, G and T, sometimes called *junk DNA* sequences, are present in the DNA of most organisms. They are also copied and passed on to the organisms' offspring. However, if such a sequence is replaced by almost any other sequence of similar length, it is still copied. So we can infer that the copying of such sequences does not depend on their specific physical form. Unlike genes, junk DNA sequences are not programs. If they have a function (and it is not known whether they do), it cannot be to carry information of any kind. Although they are copied, they do not contribute causally to their own copying, and are therefore not replicators.

Actually, that is an exaggeration. Anything that is copied must have made at least some causal contribution to that copying. Junk DNA sequences, for instance, are made of DNA, which allows the cellular computer to copy them. It cannot copy molecules other than DNA. It is not usually illuminating to consider something as a replicator if its causal contribution to its own replication is small, though strictly speaking being a replicator is a matter of degree. I shall define the *degree of adaptation* of a replicator to a given environment as the degree to which the replicator contributes causally to its own replication in that environment. If a replicator is well adapted to most environments of a niche, we may call it well adapted to the niche. We have just seen that the insulin gene is highly adapted to its niche. Junk DNA sequences have a negligible degree of adaptation by comparison with the insulin gene, or any other bona fide gene, but they are far more adapted to that niche than most molecules are.

Notice that to quantify degrees of adaptation, we have to consider not only the replicator in question but also a range of variants of it. The more sensitive the copying in a given environment is to the replicator's exact physical structure, the more adapted the replicator is to that environment. For highly adapted replicators (which are the only ones worth calling replicators) we need consider only fairly small variations, because under most large variations they would no longer be replicators. So we are contemplating replacing the replicator by broadly similar objects. To quantify the degree of adaptation to a niche, we have to consider the replicator's degree of adaptation to each environment of the niche. We must therefore consider variants of the environment as well as of the replicator. If most variants of the replicator fail to cause most environments of its niche to copy them, then it would follow that our replicator's form is a significant cause of its own copying in that niche, which is what we mean by saying that it is highly adapted to the niche. On the other hand, if most variants of the replicator would be copied in most of the environments of the niche, then the form of our replicator makes little difference, in that copying would occur anyway. In that case, our replicator makes little causal contribution to its copying, and it is not highly adapted to that niche.

So the degree of adaptation of a replicator depends not only on what that replicator does in its actual environment, but also on what a vast number of other objects, most of which do not exist, *would* do, in a vast number of environments other than the actual one. We have encountered this curious sort of property before. The accuracy of a virtual-reality rendering depends not only on the responses the machine actually makes to what the user actually does, but also on responses it does not, in the event, make to things the user does not in fact do. This similarity between living processes and virtual reality is no coincidence, as I shall shortly explain.

The most important factor determining a gene's niche is usually that the gene's replication depends on the presence of other genes. For example, the replication of a bear's insulin gene depends not only on the presence, in the bear's body, of all its other genes, but also on the presence, in the external environment, of genes from other organisms. Bears cannot survive without food, and the genes for manufacturing that food exist only in other organisms.

Different types of gene which need each other's cooperation to replicate often live joined together in long DNA chains, the DNA of an *organism*. An organism is the sort of thing – such as an animal, plant or microbe – which in everyday terms we usually think of as being alive. But it follows from what I have said that 'alive' is at best a courtesy title when applied to the parts of an organism other than its DNA. *An organism is not a replicator*: it is part of the environment of replicators – usually the most important part after the other genes. The remainder of the environment is the type of habitat that can be occupied by the organism (such as mountain tops or ocean bottoms) and the particular life-style within that habitat (such as hunter or filter-feeder) which enables the organism to survive for long enough for its genes to be replicated.

In everyday parlance we speak of organisms 'reproducing themselves'; indeed, this was one of the supposed 'characteristics of living things'. In other words, we think of organisms as replicators. But this is inaccurate. Organisms *are not copied* during reproduction; far less do they cause their own copying. They are constructed afresh according to blueprints embodied in the parent organisms' DNA. For example, if the shape of a bear's nose is altered in an accident, it may change the life-style of that particular bear, and the bear's chances of surviving to 'reproduce itself' may be affected for better or worse. But the bear with the new shape of nose has no chance of being *copied*. If it does have offspring, they will have noses of the original shape. But make a change in the corresponding gene (if you do it just after the bear is conceived, you need only change one molecule), and any offspring will not only have noses of the new shape, but copies of the new gene as well. This shows that the shape of each nose is caused by that gene, and not by the shape of any previous nose. So the shape of the bear's nose makes no causal contribution to the shape of the offspring's nose. But the shape of the bear's genes contributes both to their own copying and to the shape of the bear's nose *and* of its offspring's nose.

So an organism is the immediate environment which copies the real replicators: the organism's genes. Traditionally, a bear's nose and its den would have been classified as living and non-living entities, respectively. But that distinction is not rooted in any significant difference. The role of the bear's nose is fundamentally no different from that of its den. Neither is a replicator, though new instances of them are continually being made. Both the nose and the den are merely parts of the environment which the bear's genes manipulate in the course of getting themselves replicated.

This gene-based understanding of life – regarding organisms as part of the environment of genes – has implicitly been the basis of biology since Darwin, but it was overlooked until at least the 1960s, and not fully understood until Richard Dawkins published *The Selfish Gene* (1976) and *The Extended Phenotype* (1982).

I now return to the question whether life is a fundamental phenomenon of nature. I have warned against the reductionist assumption that emergent phenomena, such as life, are necessarily less fundamental than microscopic physical ones. Nevertheless, everything I have just been saying about what life is seems to point to its being a mere side-effect at the end of a long chain of side-effects. For it is not merely the *predictions* of biology that reduce, in principle, to those of physics: it is, on the face of it, also the explanations. As I have said, the great explanatory theories of Darwin (in modern versions such as that propounded by Dawkins), and of modern biochemistry, *are* reductive. Living molecules – genes – are merely molecules, obeying the same laws of physics and chemistry as non-living ones. They contain no special substance, nor do they have any special physical attributes. They just happen, in certain environments, to be replicators. The property of being a replicator is highly contextual – that is, it depends on intricate details of the replicator's environment: an entity is a replicator in one environment and not in another. Also, the property of being adapted to a niche does not depend on any simple, intrinsic physical attribute that the replicator has at the time, but on effects that it may cause in the future – and under hypothetical circumstances at that (i.e. in variants of the environment). Contextual and hypothetical properties are essentially derivative, so it is hard to see how a phenomenon characterized only by such properties could possibly be a fundamental phenomenon of nature.

As for the physical impact of life, the conclusion is the same: the effects of life seem negligibly small. For all we know, the planet Earth is the only place in the universe where life exists. Certainly we have seen no evidence of its existence elsewhere, so even if it is quite widespread its effects are too small to be perceptible to us. What we do see beyond the Earth is an active universe, seething with diverse, powerful but totally inanimate processes. Galaxies revolve. Stars condense, shine, flare, explode and collapse. High-energy particles and electromagnetic and gravitational waves stream in all directions. Whether life is or is not out there among all those titanic processes seems to make no difference. It seems that none of them would be in the slightest way affected if life *were* present. If the Earth were enveloped in a large solar flare, itself an insignificant event astrophysically, our biosphere would be instantly sterilized, and that catastrophe would have as little effect on the sun as a raindrop has on an erupting volcano. Our biosphere is, in terms of its mass, energy or any similar astrophysical measure of significance, a negligible fraction even of the Earth, yet it is a truism of astronomy that the solar system consists essentially of the Sun and Jupiter. Everything else (including the Earth) is 'just impurities'. Moreover, the solar system is a negligible component of our Galaxy, the Milky Way, which is itself unremarkable among the many in the known universe. So it seems that, as Stephen Hawking put it, 'The human race is just a chemical scum on a moderate-sized planet, orbiting round a very average star in the outer suburb of one among a hundred billion galaxies.'

Thus the prevailing view today is that life, far from being central, either geometrically, theoretically or practically, is of almost inconceivable insignificance. Biology, in this picture, is a subject with the same status as geography. Knowing the layout of the city of Oxford is important to those of us who live there, but unimportant to those who never visit Oxford. Similarly, it seems that life is a property of some parochial area, or perhaps areas, of the universe, fundamental to us because we are alive, but not at all fundamental either theoretically or practically in the larger scheme of things.

But remarkably, this appearance is misleading. It is simply not true that life is insignificant in its physical effects, nor is it theoretically derivative.

As a first step to explaining this, let me explain my earlier remark that life is a form of virtual-reality generation. I have used the word 'computers' for the mechanisms that execute gene programs inside living cells, but that is slightly loose terminology. Compared with the general-purpose computers that we manufacture artificially, they do more in some respects and less in others. One could not easily program them to do word processing or to factorize large numbers. On the other hand, they exert exquisitely accurate, interactive control over the responses of a complex environment (the organism) to everything that may happen to it. And this control is directed towards causing the environment to act back upon the genes in a specific way (namely, to replicate them) such that the net effect on the genes is as independent as possible of what may be happening outside. This is more than just computing. It is virtual-reality rendering.

The analogy with the human technology of virtual reality is not perfect. First, although genes are enveloped, just as a user of virtual reality is, in an environment whose detailed constitution and behaviour are specified by a program (which the genes themselves embody), the genes do not *experience* that environment because they have neither senses nor experiences. So if an organism is a virtual-reality rendering specified by its genes, it is a rendering without an audience. Second, the organism is not only being rendered, it is being manufactured. It is not a matter of 'fooling' the gene into believing that there is an organism out there. The organism really is out there.

However, these differences are unimportant. As I have said, *all* virtual-reality rendering physically manufactures the rendered environment. The inside of any virtual-reality generator in the act of rendering is precisely a real, physical environment, manufactured to have the properties specified in the program. It is just that we users sometimes choose to interpret it as a different environment, which happens to feel the same. As for the absence of a user, let us consider explicitly what the role of the user of virtual reality is. First, it is to kick the rendered environment and to be kicked back in return – in other words, to interact with the environment in an autonomous way. In the biological case, that role is performed by the external habitat. Second, it is to provide the *intention* behind the rendering. That is to say, it makes little sense to speak of a particular situation as being a virtual-reality rendering if there is no concept of the rendering being accurate or inaccurate. I have said that the accuracy of a rendering is the closeness, as perceived by the user, of the rendered environment to the intended one. But what does accuracy mean for a rendering which no one intended and no one perceives? It means the degree of adaptation of the genes to their niche. We can infer the 'intention' of genes to render an environment that will replicate them, from Darwin's theory of evolution. Genes become extinct if they do not enact that 'intention' as efficiently or resolutely as other competing genes.

So living processes and virtual-reality renderings are, superficial differences aside, the same sort of process. Both involve the physical embodying of general theories about an environment. In both cases these theories are used to realize that environment and to control, interactively, not just its instantaneous appearance but also its detailed response to general stimuli.

Genes embody knowledge about their niches. Everything of fundamental significance about the phenomenon of life depends on this property, and not on replication *per se*. So we can now take the discussion beyond replicators. In principle, one could imagine a species whose genes were unable to replicate, but instead were adapted to keep their physical form unchanged by continual self-maintenance and by protecting themselves from external influences. Such a species is unlikely to evolve naturally, but it might be constructed artificially. Just as the degree of adaptation of a replicator is defined as the degree to which it contributes causally to its own replication, we can define the degree of adaptation of these non-replicating genes as the degree to which they contribute to their own survival in a particular form. Consider a species whose genes were patterns etched in diamond. An ordinary diamond with a haphazard shape might survive for aeons under a wide range of circumstances, but that shape is not *adapted* for survival because a differently shaped diamond would also survive under similar circumstances. But if the diamond-encoded genes of our hypothetical species caused the organism to behave in a way which, for instance, protected the diamond's etched surface from corrosion in a hostile environment, or defended it against other organisms that would try to etch different information into it, or against thieves who would cut and polish it into a gemstone, then it would contain genuine adaptations for survival in those environments. (Incidentally, a gemstone *does* have a degree of adaptation for survival in the environment of present-day Earth. Humans seek out uncut diamonds and change their shapes to those of gemstones. But they seek out gemstones and preserve their shapes. So in this environment, the shape of a gemstone contributes causally to its own survival.)

When the manufacture of these artificial organisms ceased, the number of instances of each non-replicating gene could never again increase. But nor would it decrease, so long as the knowledge it contained was sufficient for it to enact its survival strategy in the niche it occupied. Eventually a sufficiently large change in the habitat, or attrition caused by accidents, might wipe out the species, but it might well survive for as long as many a naturally occurring species. The genes of such species share all the properties of real genes except replication. In particular, they embody the knowledge necessary to render their organisms in just the way that real genes do.

It is the *survival of knowledge*, and not necessarily of the gene or any other physical object, that is the common factor between replicating and non-replicating genes. So, strictly speaking, it is a piece of knowledge rather than a physical object that is or is not adapted to a certain niche. If it is adapted, then it has the property that once it is embodied in that niche, it will tend to remain so. With a replicator, the physical material that embodies it keeps changing, a new copy being assembled out of non-replicating components every time replication occurs. Non-replicating knowledge may also be successively embodied in *different* physical forms, as for example when a vintage sound recording is transferred from a vinyl record to magnetic tape, and later to compact disc. One could imagine another artificial non-replicator-based living organism that did the same sort of thing, taking every opportunity to recopy the knowledge in its genes onto the safest medium available. Perhaps one day our descendants will do that.

I think it would be perverse to call the organisms of these hypothetical species 'inanimate', but the terminology is not really important. The point is that although all known life is based on replicators, what the phenomenon of life is really about is knowledge. We can give a definition of adaptation directly in terms of knowledge: an entity is adapted to its niche if it embodies knowledge that causes the niche to keep that knowledge in existence. Now we are getting closer to the reason why life is fundamental. Life is about the physical embodiment of knowledge, and in Chapter 6 we came across a law of physics, the Turing principle, which is also about the physical embodiment of knowledge. It says that it is possible to embody the laws of physics, as they apply to every physically possible environment, in programs for a virtual-reality generator. Genes are such programs. Not only that, but all other virtual-reality programs that physically exist, or will ever exist, are direct or indirect effects of life. For example, the virtual-reality programs that run on our computers and in our brains are indirect effects of human life. So life is the means – presumably a necessary means – by which the effects referred to in the Turing principle have been implemented in nature.

This is encouraging, but it is not quite sufficient to establish that life is a fundamental phenomenon. For I have not yet established that the Turing principle itself has the status of a fundamental law. A sceptic might argue that it does not. It is a law about the physical embodiment of knowledge, and the sceptic might take the view that knowledge is a parochial, anthropocentric concept rather than a fundamental one. That is, it is one of those things which is significant to us because of what we are – animals whose ecological niche depends on creating and applying knowledge – but not significant in an absolute sense. To a koala bear, whose ecological niche depends on eucalyptus leaves, eucalyptus is significant; to the knowledge-wielding ape *Homo sapiens*, knowledge is significant.

But the sceptic would be wrong. Knowledge is significant not only to *Homo sapiens*, nor only on the planet Earth. I have said that whether something does or does not have a large physical impact is not decisive as to whether it is fundamental in nature. But it is relevant. Let us consider the astrophysical effects of knowledge.

The theory of stellar evolution – the structure and development of stars – is one of the success stories of science. (Note the clash of terminology here. The word 'evolution' in physics means development, or simply motion, not variation and selection.) Only a century ago, even the source of the Sun's energy was unknown. The best physics of the day provided only the false conclusion that whatever its energy source was, the Sun could not have been shining for more than a hundred million years. Interestingly, the geologists and palaeontologists already knew, from fossil evidence of what life had been doing, that the Sun must have been shining on the Earth for a billion years at least. Then nuclear physics was discovered, and was applied in great detail to the physics of the interiors of stars. Since then the theory of stellar evolution has matured. We now understand what makes a star shine. For most types of star we can predict what temperature, colour, luminosity and diameter it has at each stage of its history, how long each stage lasts, what elements the star creates by nuclear transmutation, and so on. This theory has been tested and borne out by observations of the Sun and other stars.

We can use the theory to predict the future development of the Sun. It says that the Sun will continue to shine with great stability for another five billion years or so; then it will expand to about a hundred times its present diameter to become a red giant star; then it will pulsate, flare into a nova, collapse and cool, eventually becoming a black dwarf. But will all this really happen to the Sun? Has *every* star that formed a few billion years before the Sun, with the same mass and composition, already become a red giant, as the theory predicts? Or is it possible that some apparently insignificant chemical processes on minor planets orbiting those stars might alter the course of nuclear and gravitational processes having overwhelmingly more mass and energy?

If the Sun does become a red giant, it will engulf and destroy the Earth. If any of our descendants, physical or intellectual, are still on the Earth at that time, they might not want that to happen. They might do everything in their power to prevent it.

Is it obvious that they will not be able to? Certainly, our present technology is far too puny to do the job. But neither our theory of stellar evolution nor any other physics we know gives any reason to believe that the task is impossible. On the contrary, we already know, in broad terms, what it would involve (namely, removing matter from the Sun). And we have several billion years to perfect our half-baked plans and put them into practice. If, in the event, our descendants do succeed in saving themselves in this way, then our present theory of stellar evolution, when applied to one particular star, the Sun, gives entirely the wrong answer. And the reason why it gives the wrong answer is that it does not take into account the effect of life on stellar evolution. It takes into account such fundamental physical effects as nuclear and electromagnetic forces, gravity, hydrostatic pressure and radiation pressure – but not life.

It seems likely that the knowledge required to control the Sun in this way could not evolve by natural selection alone, so it must specifically be *intelligent* life on whose presence the future of the Sun depends. Now, it may be objected that it is a huge and unsupported assumption that intelligence will survive on Earth for several billion years, and even if it does, it is a further assumption that it will then possess the knowledge required to control the Sun. One current view is that intelligent life on Earth is even now in danger of destroying itself, if not by nuclear war then by some catastrophic side-effect of technological advance or scientific research. Many people think that if intelligent life is to survive on Earth, it will do so only by suppressing technological progress. So they might fear that our developing the technology required to regulate stars is incompatible with surviving for long enough to use that technology, and therefore that life on Earth is destined, one way or another, not to affect the evolution of the Sun.

I am sure that this pessimism is misguided, and, as I shall explain in Chapter 14, there is every reason to conjecture that our descendants will eventually control the Sun and much more. Admittedly, we can foresee neither their technology nor their wishes. They may choose to save themselves by emigrating from the solar system, or by refrigerating the Earth, or by any number of methods, inconceivable to us, that do not involve tampering with the Sun. On the other hand, they may wish to control the Sun much sooner than would be required to prevent it from entering its red giant phase (for example to harness its energy more efficiently, or to quarry it for raw materials to construct more living space for themselves). However, the point I am making here does not depend on our being able to predict what will happen, but only on the proposition that what will happen will depend on what knowledge our descendants have, and on how they choose to apply it. Thus one cannot predict the future of the Sun without taking a position on the future of life on Earth, and in particular on the future of knowledge. The colour of the Sun ten billion years hence depends on gravity and radiation pressure, on convection and nucleosynthesis. It does not depend at all on the geology of Venus, the chemistry of Jupiter, or the pattern of craters on the Moon. But it does depend on what happens to intelligent life on the planet Earth. It depends on politics and economics and the outcomes of wars. It depends on what people do: what decisions they make, what problems they solve, what values they adopt, and on how they behave towards their children.

One cannot avoid this conclusion by adopting a pessimistic theory of the prospects for our survival. Such a theory does not follow from the laws of physics or from any other fundamental principle that we know, and can be justified only in high-level, human terms (such as 'scientific knowledge has outrun moral knowledge', or whatever). So, in arguing from such a theory one is implicitly conceding that theories of human affairs are necessary for making astrophysical predictions. And even if the human race will in the event fail in its efforts to survive, does the pessimistic theory apply to every extraterrestrial intelligence in the universe? If not – if some intelligent life, in some galaxy, will ever succeed in surviving for billions of years – then life is significant in the gross physical development of the universe.

Throughout our Galaxy and the multiverse, stellar evolution depends on whether and where intelligent life has evolved, and if so, on the outcomes of *its* wars and on how it treats *its* children. For example, we can predict roughly what proportions of stars of different colours (more precisely, of different spectral types) there should be in the Galaxy. To do that we shall have to make some assumptions about how much intelligent life there is out there, and what it has been doing (namely, that it has not been switching off too many stars). At the moment, our observations are consistent with there being no intelligent life outside our solar system. When our theories of the structure of our Galaxy are further refined, we shall be able to make more precise predictions, but again only on the basis of assumptions about the distribution and behaviour of intelligence in the Galaxy. If those assumptions are inaccurate we shall predict the wrong distribution of spectral types just as surely as if we were to make a mistake about the composition of interstellar gases, or about the mass of the hydrogen atom. And, if we detect certain anomalies in the distribution of spectral types, this could be evidence of the presence of extraterrestrial intelligence.

The cosmologists John Barrow and Frank Tipler have considered the astrophysical effects that life would have if it survived for long *after* the time at which the Sun would otherwise become a red giant. They have found that life would eventually make major, qualitative changes to the structure of the Galaxy, and later to the structure of the whole universe. (I shall return to these results in Chapter 14.) So once again, any theory of the structure of the universe in all but its earliest stages must take a position on what life will or will not be doing by then. There is no getting away from it: the future history of the universe depends on the future history of knowledge. Astrologers used to believe that cosmic events influence human affairs; science believed for centuries that neither influences the other. Now we see that human affairs influence cosmic events.

It is worth reflecting on where we went astray in underestimating the physical impact of life. It was by being too parochial. (That is ironic, because the ancient consensus happened to avoid our mistake by being even more parochial.) In the universe *as we see it*, life has affected nothing of any astrophysical significance. However, we see only the past, and it is only the past of what is spatially near us that we see in any detail. The further we look into the universe, the further back in time we see and the less detail we see. But even the whole past – the history of the universe from the Big Bang until now – is just a small part of physical reality. There is at least ten times as much history still to go, between now and the Big Crunch (if that happens), and probably a lot more, to say nothing of the other universes. We cannot observe any of this, but when we apply our best theories to the future of the stars, and of the galaxies and the universe, we find plenty of scope for life to affect and, in the long run, to dominate everything that happens, just as it now dominates the Earth's biosphere.

The conventional argument for the insignificance of life gives too much weight to bulk quantities like size, mass and energy. In the parochial past and present these were and are good measures of astrophysical significance, but there is no reason within physics why that should continue to be so. Moreover, the biosphere itself already provides abundant counter-examples to the general applicability of such measures of significance. In the third century BC, for instance, the mass of the human race was about ten million tonnes. One might therefore conclude that it is unlikely that physical processes occurring in the third century BC and involving the motion of many times that mass could have been significantly affected by the presence or absence of human beings. But the Great Wall of China, whose mass is about three hundred million tonnes, was built at that time. Moving millions of tonnes of rock is the sort of thing that human beings do all the time. Nowadays it takes only a few dozen humans to excavate a million-tonne railway cutting or tunnel. (The point is made even more strongly if we make a fairer comparison, between the mass of rock shifted and the mass of that tiny part of the engineer's, or emperor's, brain that embodies the ideas, or memes, that cause the rock to be shifted.) The human race as a whole (or, if you like, its stock of memes) probably already has enough knowledge to destroy whole planets, if its survival depended on doing so. Even non-intelligent life has grossly transformed many times its own mass of the surface and atmosphere of the Earth. All the oxygen in our atmosphere, for instance – about a thousand trillion tonnes – was created by plants and was therefore a side-effect of the replication of genes, i.e. molecules, which were descendants of a single molecule. Life achieves its effects not by being larger, more massive or more energetic than other physical processes, but by being more knowledgeable. In terms of its gross effect on the outcomes of physical processes, knowledge is at least as significant as any other physical quantity.

But is there, as the ancients assumed there must be in the case of life, a basic physical difference between knowledge-bearing and non-knowledge-bearing objects, a difference that depends neither on the objects' environments nor on their effects on the remote future, but only on the objects' immediate physical attributes? Remarkably, there is. To see what it is, we must take the multiverse view.

Consider the DNA of a living organism, such as a bear, and suppose that somewhere in one of its genes we find the sequence TCGTCGTTTC. That particular string of ten molecules, in the special niche consisting of the rest of the gene and *its* niche, is a replicator. It embodies a small but significant amount of knowledge. Now suppose, for the sake of argument, that we can find a junk-DNA (non-gene) segment in the bear's DNA which also has the sequence TCGTCGTTTC. Nevertheless this sequence is not worth calling a replicator, because it contributes almost nothing to its replication, and it embodies no knowledge. It is a random sequence. So here we have two physical objects, both segments of the same DNA chain, one of which embodies knowledge and the other is a random sequence. But they are *physically identical*. How can knowledge be a fundamental physical quantity, if one object has it while a physically identical object does not?

It can, because these two segments are not really identical. They only look identical when viewed from some universes, such as ours. Let us look at them again, as they appear in other universes. We cannot directly observe other universes, so we must use theory.

We know that DNA in living organisms is naturally subject to random variations – *mutations* – in the sequence of A, C, G and T molecules. According to the theory of evolution, the adaptations in genes, and therefore the genes' very existence, depend on such mutations having occurred. Because of mutations, populations of any gene contain a degree of variation, and individuals carrying genes with higher degrees of adaptation tend to have more offspring than other individuals. Most variations in a gene make it unable to cause its replication, because the altered sequence no longer instructs the cell to manufacture anything useful. Others merely make replication less likely (that is, they narrow the gene's niche). But some may happen to embody new instructions that make replication *more* likely. Thus natural selection occurs. With each generation of variation and replication the degree of adaptation of the surviving genes tends to increase. Now, a random mutation, caused for instance by a cosmic-ray strike, causes variation not only within the population of the organism in one universe, but between universes as well. A cosmic 'ray' is a high-energy sub-atomic particle, and, like a photon emitted from a torch, it travels in different directions in different universes. So when a cosmic-ray particle strikes a DNA strand and causes a mutation, some of its counterparts in other universes are missing their copies of the DNA strand altogether, while others are striking it at different positions, and hence causing different mutations. Thus a single cosmic-ray strike on a single DNA molecule will in general cause a large range of different mutations to appear in different universes.

When we are considering what a particular object may look like in other universes, we must not look so far afield in the multiverse that it is impossible to identify a counterpart, in the other universe, of that object. Take a DNA segment, for instance. In some universes there are no DNA molecules at all. Some universes containing DNA are so dissimilar to ours that there is no way of identifying which DNA segment in the other universe corresponds to the one we are considering in this universe. It is meaningless to ask what our particular DNA segment looks like in such a universe, so we must consider only universes that are sufficiently similar to ours for this ambiguity not to arise. For instance, we could consider only those universes in which bears exist, and in which a sample of DNA from a bear has been placed in an analysing machine, which has been programmed to print out ten letters representing the structure at a specified position relative to certain landmarks on a specified DNA strand. The following discussion would be unaffected if we were to choose any other reasonable criterion for identifying corresponding segments of DNA in nearby universes.

By any such criterion, the bear's gene segment must have the same sequence in almost all nearby universes as it does in ours. That is because it is presumably highly adapted, which means that most variants of it would not succeed in getting themselves copied in most variants of their environment, and so could not appear at that location in the DNA of a living bear. In contrast, when the non-knowledge-bearing DNA segment undergoes almost any mutation, the mutated version is still capable of being copied. Over generations of replication many mutations will have occurred, and most of them will have had no effect on replication. Therefore the junk-DNA segment, unlike its counterpart in the gene, will be thoroughly heterogeneous in different universes. It may well be that every possible variation of its sequence is equally represented in the multiverse (that is what we should mean by its sequence being strictly random).

So the multiverse perspective reveals additional physical structure in the bear's DNA. In this universe, it contains two segments with the sequence TCGTCGTTTC. One of them is part of a gene, while the other is not part of any gene. In most other nearby universes, the first of the two segments has the same sequence, TCGTCGTTTC, as it does in our universe, but the second segment varies greatly between nearby universes. So from the multiverse perspective the two segments are not even remotely alike (Figure 8.1).

Again we were too parochial, and were led to the false conclusion that knowledge-bearing entities can be physically identical to non-knowledge-bearing ones; and this in turn cast doubt on the fundamental status of knowledge. But now we have come almost full circle. We can see that the ancient idea that living matter has special physical properties was almost true: it is not living matter but *knowledge-bearing* matter that is physically special. Within one universe it looks irregular; across universes it has a regular structure, like a crystal in the multiverse.

So knowledge is a fundamental physical quantity after all, and the phenomenon of life is only slightly less so.

Imagine looking through an electron microscope at a DNA molecule from a bear's cell, and trying to distinguish the genes from the non-gene sequences and to estimate the degree of adaptation of each gene. In any one universe, this task is impossible. The

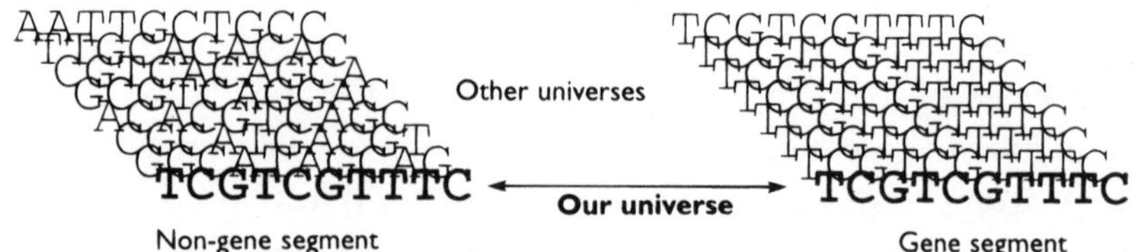

FIGURE 8.1 *Multiverse view of two DNA segments which happen to be identical in our universe, one random and one from within a gene.*

property of being a gene – that is, of being highly adapted – is, in so far as it can be detected within one universe, overwhelmingly complicated. It is an emergent property. You would have to make many copies of the DNA, with variations, use genetic engineering to create many bear embryos for each variant of the DNA, allow the bears to grow up and live in a variety of environments representative of the bear's niche, and see which bears succeed in having offspring.

But with a magic microscope that could see into other universes (which, I stress, is not possible: we are using theory to imagine – or render – what we know must be there) the task would be easy. As in Figure 8.1, the genes would stand out from the non-genes just as cultivated fields stand out from a jungle in an aerial photograph, or like crystals that have precipitated from solution. They are regular across many nearby universes, while all the non-gene, junk-DNA segments are irregular. As for the degree of adaptation of a gene, this is almost as easy to estimate. The better-adapted genes will have the same structure over a wider range of universes – they will have bigger 'crystals'.

Now go to an alien planet, and try to find the local life-forms, if any. Again, this is a notoriously difficult task. You would have to perform complex and subtle experiments whose infinite pitfalls have been the subject of many a science-fiction story. But if only you could observe through a multiverse telescope, life and its consequences would be obvious at a glance. You need only look for complex structures that seem irregular in any one universe, but are identical across many nearby universes. If you see any, you will have found some physically embodied knowledge. Where there is knowledge, there must have been life, at least in the past.

Compare a living bear with the Great Bear constellation. The living bear is anatomically very similar in many nearby universes. It is not only its genes that have that property, but its whole body (though other attributes of its body, such as its weight, vary much more than the genes; that is because, for example, in different universes the bear has been more or less successful in its recent search for food). But in the Great Bear constellation there is no such regularity from one universe to another. The shape of the constellation is a result of the initial conditions in the galactic gas from which the stars formed. Those conditions were random – very diverse in different universes, at a microscopic level – and the process of the formation of stars from that gas involved various instabilities which amplified the scale of the variations. As a result, the pattern of stars that we see in the constellation exists in only a very narrow range of universes. In most nearby variants of our universe there are also constellations in the sky, but they look different.

Finally, let us look around the universe in a similar way. What will catch our magically enhanced eye? In a single universe the most striking structures are galaxies and clusters of galaxies. But those objects have no discernible structure across the multiverse. Where there is a galaxy in one universe, a myriad galaxies with quite different geographies are stacked in the multiverse. And so it is everywhere in the multiverse. Nearby universes are alike only in certain gross features, as required by the laws of physics, which apply to them all. Thus most stars are quite accurately spherical everywhere in the multiverse, and most galaxies are spiral or elliptical. But nothing extends far into other universes without its detailed structure changing unrecognizably. Except, that is, in those few places where there is embodied knowledge. In such places, objects extend recognizably across large numbers of universes. Perhaps the Earth is the only such place in our universe, at present. In any case, such places stand out, in the sense I have described, as the location of the processes – life, and thought – that have generated the largest distinctive structures in the multiverse.

TERMINOLOGY

replicator An entity that causes certain environments to make copies of it.

gene A molecular replicator. Life on Earth is based on genes that are DNA strands (RNA in the case of some viruses).

meme An idea that is a replicator, such as a joke or a scientific theory.

niche The niche of a replicator is the set of all possible environments in which the replicator would cause its own replication. The niche of an organism is the set of all possible environments and life-styles in which it could live and reproduce.

adaptation The degree to which a replicator is adapted to a niche is the degree to which it causes its own replication in that niche. More generally, an entity is adapted to its niche to the extent that it embodies knowledge that causes the niche to keep that knowledge in existence.

SUMMARY

Scientific progress since Galileo has seemed to refute the ancient idea that life is a fundamental phenomenon of nature. It has revealed the vast scale of the universe, compared with the Earth's biosphere. Modern biology seems to have confirmed this refutation, by explaining living processes in terms of molecular replicators, genes, whose behaviour is governed by the same laws of physics as apply to inanimate matter. Nevertheless, life *is* associated with a fundamental principle of physics – the Turing principle – since it is the means by which virtual reality was first realized in nature. Also, despite appearances, life *is* a significant process on the largest scales of both time and space. The future behaviour of life will determine the future behaviour of stars and galaxies. And the largest-scale regular structure *across* universes exists where knowledge-bearing matter, such as brains or DNA gene segments, has evolved.

This direct connection between the theory of evolution and quantum theory is, to my mind, one of the most striking and unexpected of the many connections between the four strands. Another is the existence of a substantive quantum theory of computation *underlying the existing theory of computation. That connection is the subject of the next chapter.*

9

Quantum Computers

To anyone new to the subject, *quantum computation* sounds like the name of a new technology – the latest, perhaps, in the remarkable succession that has included mechanical computation, transistorized electronic computation, silicon-chip computation, and so on. And it is true that even existing computer technology relies on microscopic quantum-mechanical processes. (Of course *all* physical processes are quantum-mechanical, but here I mean ones for which classical physics – i.e. non-quantum physics – gives very inaccurate predictions.) If the trend towards ever faster, more compact computer hardware is to continue, the technology must become even more 'quantum-mechanical' in this sense, simply because quantum-mechanical effects are dominant in all sufficiently small systems. If there were no more to it than that, quantum computation could hardly figure in any fundamental explanation of the fabric of reality, for there would be nothing fundamentally new in it. All present-day computers, whatever quantum-mechanical processes they may exploit, are merely different technological implementations of the same *classical* idea, that of the universal Turing machine. That is why the repertoire of computations available to all existing computers is essentially the same: they differ only in their speed, memory capacity and input–output devices. That is to say, even the lowliest of today's home computers can be programmed to solve any problem, or render any environment, that our most powerful computers can, provided only that it is given additional memory, allowed to run for long enough, and given appropriate hardware for displaying its results.

Quantum computation is more than just a faster, more miniaturized technology for implementing Turing machines. A *quantum computer* is a machine that uses uniquely quantum-mechanical effects, especially interference, to perform wholly new types of computation that would be impossible, even in principle, on any Turing machine and hence on any classical computer. Quantum computation is therefore nothing less than a distinctively new way of harnessing nature.

Let me elaborate that claim. The earliest inventions for harnessing nature were tools powered by human muscles. They revolutionized our ancestors' situation, but they suffered from the limitation that they required continuous human attention and effort during every moment of their use. Subsequent technology overcame that limitation: human beings managed to domesticate certain animals and plants, turning the biological adaptations in those organisms to human ends. Thus the crops could grow, and the guard dogs could watch, even while their owners slept. Another new type of technology began when human beings went beyond merely exploiting existing adaptations (and existing non-biological phenomena such as fire), and created completely new adaptations in the world, in the form of pottery, bricks, wheels, metal artefacts and machines. To do this they had to think about, and understand, the natural laws governing the world – including, as I have explained, not only its superficial aspects but the underlying fabric of reality. There followed thousands of years of progress in this type of technology – harnessing some of the *materials, forces* and *energies* of physics. In the twentieth century *information* was added to this list when the invention of computers allowed complex information processing to be performed outside human brains. *Quantum* computation, which is now in its early infancy, is a distinct further step in this progression. It will be the first technology that allows useful tasks to be performed in collaboration between parallel universes. A quantum computer would be capable of distributing components of a complex task among vast numbers of parallel universes, and then sharing the results.

I have already mentioned the significance of computational

universality – the fact that a single physically possible computer can, given enough time and memory, perform any computation that any other physically possible computer can perform. The laws of physics as we currently know them do admit computational universality. However, to be at all useful or significant in the overall scheme of things, universality as I have defined it up to now is not sufficient. It merely means that the universal computer can *eventually* do what any other computer can. In other words, *given enough time* it is universal. But what if it is not given enough time? Imagine a universal computer that could execute only one computational step in the whole lifetime of the universe. Would *its* universality still be a profound property of reality? Presumably not. To put that more generally, one can criticize this narrow notion of universality because it classifies a task as being in a computer's repertoire regardless of the physical resources that the computer would expend in performing the task. Thus, for instance, we have considered a virtual-reality user who is prepared to go into suspended animation for billions of years, while the computer calculates what to show next. In discussing the ultimate limits of virtual reality, that is the appropriate attitude for us to take. But when we are considering the *usefulness* of virtual reality – or what is even more important, the fundamental role that it plays in the fabric of reality – we must be more discriminating. Evolution would never have got off the ground if the task of rendering certain properties of the earliest, simplest habitats had not been *tractable* (that is, computable in a reasonable time) using readily available molecules as computers. Likewise, science and technology would never have got off the ground if designing a stone tool had required a thousand years of thinking. Moreover, what was true at the beginning has remained an absolute condition for progress at every step. Computational universality would not be much use to genes, no matter how much knowledge they contained, if rendering their organism were an intractable task – say, if one reproductive cycle took billions of years.

Thus the fact that there *are* complex organisms, and that there *has* been a succession of gradually improving inventions and scientific theories (such as Galilean mechanics, Newtonian mechanics, Einsteinian mechanics, quantum mechanics, . . .) tells us something more about what sort of computational universality exists in reality. It tells us that the actual laws of physics are, thus far at least, capable of being successively approximated by theories that give ever better explanations and predictions, and that the task of discovering each theory, given the previous one, has been computationally tractable, given the previously known laws and the previously available technology. The fabric of reality must be, as it were, *layered*, for easy self-access. Likewise, if we think of evolution itself as a computation, it tells us that there have been sufficiently many viable organisms, coded for by DNA, to allow better-adapted ones to be computed (i.e. to evolve) using the resources provided by their worse-adapted predecessors. So we can infer that the laws of physics, in addition to mandating their own comprehensibility through the Turing principle, ensure that the corresponding evolutionary processes, such as life and thought, are neither too time-consuming nor require too many resources of any other kind to occur in reality.

So, the laws of physics not only permit (or, as I have argued, *require*) the existence of life and thought, they require them to be, in some appropriate sense, efficient. To express this crucial property of reality, modern analyses of universality usually postulate computers that are universal in an even stronger sense than the Turing principle would, on the face of it, require: not only are universal virtual-reality generators possible, it is possible to build them so that they do not require impracticably large resources to render simple aspects of reality. From now on, when I refer to universality I shall mean it in this sense, unless otherwise stated.

Just how efficiently can given aspects of reality be rendered? What computations, in other words, are practicable in a given time and under a given budget? This is the basic question of computational complexity theory which, as I have said, is the study of the resources that are required to perform given computational tasks. Complexity theory has not yet been sufficiently well integrated with physics to give many quantitative answers. However, it has made a fair amount of headway in defining a useful, rough-and-ready distinction between *tractable* and *intractable* computational tasks. The general approach is best illustrated by an example. Consider the task of multiplying together two rather large numbers, say 4,220,851 and 2,594,209. Many of us remember the method we learned in childhood for performing such multiplications. It involves multiplying each digit of one number in turn by each digit of the other, while shifting and adding the results together in a standard way to give the final answer, in this case 10,949,769,651,859. Many might be loath to concede that this wearisome procedure makes multiplication 'tractable' in any ordinary sense of the word. (Actually there are more efficient methods for multiplying large numbers, but this one provides a good enough illustration.) But from the point of view of complexity theory, which deals in massive tasks carried out by computers that are not subject to boredom and almost never make mistakes, this method certainly does fall into the 'tractable' category.

What counts for 'tractability', according to the standard definitions, is not the actual time taken to multiply a particular pair of numbers, but the fact that the time does not increase too sharply when we apply the same method to ever larger numbers. Perhaps surprisingly, this rather indirect way of defining tractability works very well in practice for many (though not all) important classes of computational tasks. For example, with multiplication we can easily see that the standard method can be used to multiply numbers that are, say, about ten times as large, with very little extra work. Suppose, for the sake of argument, that each elementary multiplication of one digit by another takes a certain computer one microsecond (including the time taken to perform the additions, shifts and other operations that follow each elementary multiplication). When we are multiplying the seven-digit numbers 4,220,851 and 2,594,209, each of the seven digits in 4,220,851 has to be multiplied by each of the seven digits in 2,594,209. So the total time required for the multiplication (if the operations are performed sequentially) will be seven times seven, or 49 microseconds. For inputs roughly ten times as large as these, which would have eight digits each, the time required to multiply them would be 64 microseconds, an increase of only 31 per cent.

Clearly, numbers over a huge range – certainly including any numbers that have ever been measured as the values of physical variables – can be multiplied in a tiny fraction of a second. So multiplication is indeed tractable for all purposes within physics (or, at least, within existing physics). Admittedly, practical reasons for multiplying much larger numbers can arise outside physics. For instance, products of prime numbers of 125 digits or so are of great interest to cryptographers. Our hypothetical machine could multiply two such prime numbers together, making a 250-digit product, in just over a hundredth of a second. In one second it could multiply two 1000-digit numbers, and real computers available today can easily improve upon those timings. Only a few researchers in esoteric branches of pure mathematics are interested in performing such incomprehensibly vast multiplications, yet we see that even they have no reason to regard multiplication as intractable.

By contrast, *factorization*, essentially the reverse of multiplication, seems much more difficult. One starts with a single number as input, say 10,949,769,651,859, and the task is to find two factors – smaller numbers which when multiplied together make 10,949,769,651,859. Since we have just multiplied them, we know that the answer in this case is 4,220,851 and 2,594,209 (and since those are both primes, it is the only correct answer). But without such inside knowledge, how would we have found the factors? You will search your childhood memories in vain for an easy method, for there isn't one.

The most obvious method of factorization is to divide the input number by all possible factors, starting with 2 and continuing with every odd number, until one of them divides the input exactly. At least one of the factors (assuming the input is not a prime) can be no larger than the input's square root, and that provides an estimate of how long the method might take. In the case we are considering, our computer would find the smaller of the two factors, 2,594,209, in just over a second. However, an input ten times as large would have a square root that was about three times as large, so factorizing it by this method would take up to three times as long. In other words, adding one digit to the input would now *triple* the running time. Adding another would triple it again, and so on. So the running time would increase in geometrical proportion, that is, exponentially, with the number of digits in the number we are factorizing. Factorizing a number with 25-digit factors by this method would occupy all the computers on Earth for centuries.

The method can be improved upon, but *all* methods of factorization currently in use have this exponential-increase property. The largest number that has been factorized 'in anger', as it were – a number whose factors were secretly chosen by mathematicians in order to present a challenge to other mathematicians – had 129 digits. The factorization was achieved, after an appeal on the Internet, by a global cooperative effort involving thousands of computers. The computer scientist Donald Knuth has estimated that the factorization of a 250-digit number, using the most efficient known methods, would take over a million years on a network of a million computers. Such things are difficult to estimate, but even if Knuth is being too pessimistic one need only consider numbers with a few more digits and the task will be made many times harder. This is what we mean by saying that the factorization of large numbers is intractable. All this is a far cry from multiplication where, as we have seen, the task of multiplying a pair of 250-digit numbers is a triviality on anyone's home computer. No one can even conceive of how one might factorize thousand-digit numbers, or million-digit numbers.

At least, no one *could* conceive of it, until recently.

In 1982 the physicist Richard Feynman considered the computer simulation of quantum-mechanical objects. His starting-point was something that had already been known for some time without its significance being appreciated, namely that predicting the behaviour of quantum-mechanical systems (or, as we can describe it, rendering quantum-mechanical environments in virtual reality) is in general an intractable task. One reason why the significance of this had not been appreciated is that no one expected the computer prediction of interesting physical phenomena to be especially easy. Take weather forecasting or earthquake prediction, for instance. Although the relevant equations are known, the difficulty of applying them in realistic situations is notorious. This has recently been brought to public attention in popular books and articles on *chaos* and the 'butterfly effect'. These effects are not responsible for the intractability that Feynman had in mind, for the simple reason that they occur only in classical physics – that is, not in reality, since reality is quantum-mechanical. Nevertheless, I want to make some remarks here about 'chaotic' classical motions, if only to highlight the quite different characters of classical and quantum unpredictability.

Chaos theory is about limitations on predictability in classical physics, stemming from the fact that almost all classical systems are inherently unstable. The 'instability' in question has nothing to do with any tendency to behave violently or disintegrate. It is about an extreme sensitivity to initial conditions. Suppose that we know the present state of some physical system, such as a set of billiard balls rolling on a table. If the system obeyed classical physics, as it does to a good approximation, we should then be able to determine its future behaviour – say, whether a particular ball will go into a pocket or not – from the relevant laws of motion, just as we can predict an eclipse or a planetary conjunction from the same laws. But in practice we are never able to measure the initial positions and velocities perfectly. So the question arises, if we know them to some reasonable degree of accuracy, can we also predict to a reasonable degree of accuracy how they will behave in the future? And the answer is, usually, that we cannot. The difference between the real trajectory and the predicted trajectory, calculated from slightly inaccurate data, tends to grow exponentially and irregularly ('chaotically') with time, so that after a while the original, slightly imperfectly known state is no guide at all to what the system is doing. The implication for computer prediction is that planetary motions, the epitome of classical predictability, are untypical classical systems. In order to predict what a typical classical system will do after only a moderate period, one would have to determine its initial state to an impossibly high precision. Thus it is said that in principle, the flap of a butterfly's wing in one hemisphere of the planet could cause a hurricane in the other hemisphere. The infeasibility of weather forecasting and the like is then attributed to the impossibility of accounting for every butterfly on the planet.

However, real hurricanes and real butterflies obey quantum theory, not classical mechanics. The instability that would rapidly amplify slight mis-specifications of an initial classical state is simply not a feature of quantum-mechanical systems. In quantum mechanics, small deviations from a specified initial state tend to cause only small deviations from the predicted final state. Instead, accurate prediction is made difficult by quite a different effect.

The laws of quantum mechanics require an object that is initially at a given position (in all universes) to 'spread out' in the multiverse sense. For instance, a photon and its other-universe counterparts all start from the same point on a glowing filament, but then move in trillions of different directions. When we later make a measurement of what has happened, we too become differentiated as each copy of us sees what has happened in our particular universe. If the object in question is the Earth's atmosphere, then a hurricane may have occurred in 30 per cent of universes, say, and not in the remaining 70 per cent. Subjectively we perceive this as a single, unpredictable or 'random' outcome, though from the multiverse point of view all the outcomes have actually happened. This parallel-universe multiplicity is the real reason for the unpredictability of the weather. Our inability to measure the initial conditions accurately is completely irrelevant. Even if we knew the initial conditions perfectly, the multiplicity, and therefore the unpredictability of the motion, would remain. And on the other hand, in contrast to the classical case, an imaginary multiverse with only slightly different initial conditions would not behave very differently from the real multiverse: it might suffer hurricanes in 30.000 001 per cent of its universes and not in the remaining 69.999 999 per cent.

The flapping of butterflies' wings does not, in reality, cause hurricanes because the classical phenomenon of chaos depends on perfect determinism, which does not hold in any single universe. Consider a group of identical universes at an instant at which, in all of them, a particular butterfly's wings have flapped up. Consider a second group of universes which at the same instant are identical to the first group, except that in them the butterfly's wings are down. Wait for a few hours. Quantum mechanics predicts that, unless there are exceptional circumstances (such as someone watching the butterfly and pressing a button to detonate a nuclear bomb if it flaps its wings), the two groups of universes, nearly identical at first, are still nearly identical. But each group, within itself, has become greatly differentiated. It includes universes with hurricanes, universes without hurricanes, and even a very tiny number of universes in which the butterfly has spontaneously changed its species through an accidental rearrangement of all its atoms, or the Sun has exploded because all its atoms bounced by chance towards the nuclear reaction at its core. Even so, the two groups still resemble each other very closely. In the universes in which the butterfly raised its wings and hurricanes occurred, those hurricanes were indeed unpredictable; but the butterfly was not causally responsible, for there were near-identical hurricanes in universes where everything else was the same but the wings were lowered.

It is perhaps worth stressing the distinction between *unpredictability* and *intractability*. Unpredictability has nothing to do with the available computational resources. Classical systems are unpredictable (or would be, if they existed) because of their sensitivity to initial conditions. Quantum systems do not have that sensitivity, but are unpredictable because they behave differently in different universes, and so appear random in most universes. In neither case will any amount of computation lessen the unpredictability. Intractability, by contrast, is a computational-resource issue. It refers to a situation where we could readily make the prediction if only we could perform the required computation, but we cannot do so because the resources required are impractically large. In order to disentangle the problems of unpredictability from those of intractability in quantum mechanics, we have to consider quantum systems that are, in principle, predictable.

Quantum theory is often presented as making only probabilistic predictions. For example, in the perforated-barrier-and-screen type of interference experiment described in Chapter 2, the photon can be observed to arrive anywhere in the 'bright' part of the shadow pattern. But it is important to understand that for many other experiments quantum theory predicts a single, definite outcome. In other words, it predicts that all universes will end up with the same outcome, even if the universes differed at intermediate stages of the experiment, and it predicts what that outcome will be. In such cases we observe *non-random interference phenomena*. An interferometer can demonstrate such phenomena. This is an optical instrument that consists mainly of mirrors, both conventional mirrors (Figure 9.1) and semi-silvered mirrors (as used in conjuring tricks

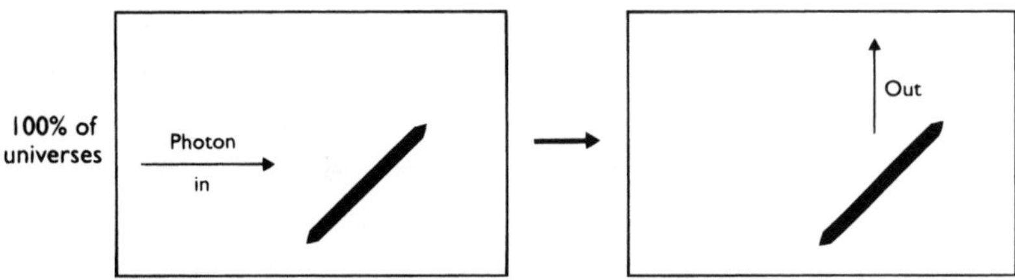

FIGURE 9.1 *The action of a normal mirror is the same in all universes.*

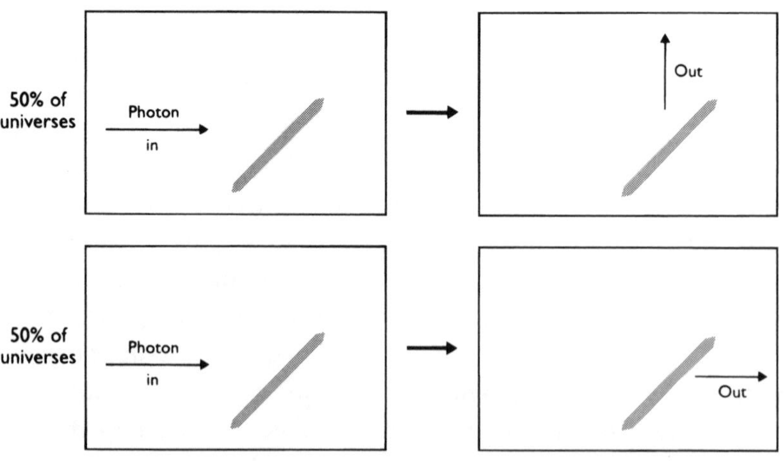

FIGURE 9.2 *A semi-silvered mirror makes initially identical universes differentiate into two equal groups, differing only in the path taken by a single photon.*

and police stations and shown in Figure 9.2). If a photon strikes a semi-silvered mirror, then in half the universes it bounces off just as it would from a conventional mirror. But in the other half, it passes through as if nothing were there.

A single photon enters the interferometer at the top left, as shown in Figure 9.3. In all the universes in which the experiment is done, the photon and its counterparts are travelling towards the interferometer along the same path. These universes are therefore identical. But as soon as the photon strikes the semi-silvered mirror, the initially identical universes become differentiated. In half of them, the photon passes straight through and travels along the top side of the interferometer. In the remaining universes, it bounces off the mirror and travels down the left side of the interferometer. The versions of the photon in these two groups of universes then strike and bounce off the ordinary mirrors at the top right and bottom left respectively. Thus they end up arriving simultaneously at the semi-silvered mirror on the bottom right, and interfere with one another. Remember that we have allowed only one photon into the apparatus, and in each universe there is still only one photon in there. In all universes, that photon has now struck the bottom-right mirror. In half of them it has struck it from the left, and in the other half it has struck it from above. The versions of the photon

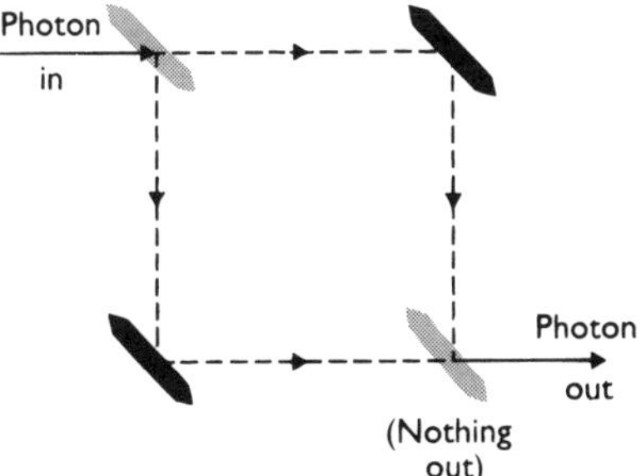

FIGURE 9.3 *A single photon passing through an interferometer. The positions of the mirrors (conventional mirrors shown black, semi-silvered mirrors grey) can be adjusted so that interference between two versions of the photon (in different universes) makes both versions take the same exit route from the lower semi-silvered mirror.*

in these two groups of universes interfere strongly. The net effect depends on the exact geometry of the situation, but Figure 9.3 shows the case where in all universes the photon ends up taking the rightward-pointing path through the mirror, and in no universe is it transmitted or reflected downwards. Thus all the universes are identical at the end of the experiment, just as they were at the beginning. They were differentiated, and interfered with one another, only for a minute fraction of a second in between.

This remarkable non-random interference phenomenon is just as inescapable a piece of evidence for the existence of the multiverse as is the phenomenon of shadows. For the outcome that I have described is incompatible with *either* of the two possible paths that a particle in a single universe might have taken. If we project a photon rightwards along the lower arm of the interferometer, for instance, it *may* pass through the semi-silvered mirror like the photon in the interference experiment does. But it may not – sometimes it is deflected downwards. Likewise, a photon projected downwards along the right arm may be deflected rightwards, as in the interference experiment, or it may just travel straight down. Thus, whichever path you set a single photon on *inside* the apparatus, it will emerge randomly. Only when interference occurs between the two paths is the outcome predictable. It follows that what is present in the apparatus just before the end of the interference experiment cannot be a single photon on a single path: it cannot, for instance, be just a photon travelling on the lower arm. There must be something else present, preventing it from bouncing downwards. Nor can there be just a photon travelling on the right arm; again, something else must be there, preventing it from travelling straight down, as it sometimes would if it were there by itself. Just as with shadows, we can construct further experiments to show that the 'something else' has all the properties of a photon that travels along the other path and interferes with the photon we see, but with nothing else in our universe.

Since there are only two different kinds of universe in this experiment, the calculation of what will happen takes only about twice as long as it would if the particle obeyed classical laws – say, if we were computing the path of a billard ball. A factor of two will hardly make such computations intractable. However, we have already seen that multiplicity of a much larger degree is fairly easy to achieve. In the shadow experiments, a single photon passes through a barrier in which there are some small holes, and then falls on a screen. Suppose that there are a thousand holes in the barrier. There are places on the screen where the photon can fall (*does* fall, in some universes), and places where it cannot fall. To calculate whether a particular point on the screen can or cannot ever receive the photon, we must calculate the mutual interference effects of a thousand parallel-universe versions of the photon. Specifically, we have to calculate one thousand paths from the barrier to the given point on the screen, and then calculate the effects of those photons on each other so as to determine whether or not they are all prevented from reaching that point. Thus we must perform roughly a thousand times as much computation as we would if we were working out whether a classical particle would strike the specified point or not.

The complexity of this sort of computation shows us that there is a lot more happening in a quantum-mechanical environment than – literally – meets the eye. And I have argued, expressing Dr Johnson's criterion for reality in terms of computational complexity, that this complexity is the core reason why it does not make sense to deny the existence of the rest of the multiverse. But far higher multiplicities are possible when there are two or more interacting particles involved in an interference phenomenon. Suppose that each of two interacting particles has (say) a thousand paths open to it. The pair can then be in a million different states at an intermediate stage of the experiment, so there can be up to a million universes that differ in what this *pair* of particles is doing. If three particles were interacting, the number of different universes could be a billion; for four, a trillion; and so on. Thus the number of different histories that we have to calculate if we want to predict what will happen in such cases increases exponentially with the number of interacting particles. That is why the task of computing how a typical quantum system will behave is well and truly intractable.

This is the intractability that was exercising Feynman. We see that it has nothing to do with unpredictability: on the contrary, it is most clearly manifested in quantum phenomena that are highly predictable. That is because in such phenomena the same, definite outcome occurs in all universes, but that outcome is the result of interference between vast numbers of universes that were different during the experiment. All this is in principle predictable from quantum theory and is not overly sensitive to the initial conditions. What makes it hard to *predict* that in such experiments the outcome will always be the same is that doing so requires inordinately large amounts of computation.

Intractability is in principle a greater impediment to universality than unpredictability could ever be. I have already said that a perfectly accurate rendering of a roulette wheel need not – indeed should not – give the same sequence of numbers as the real one. Similarly, we cannot prepare in advance a virtual-reality rendering of tomorrow's weather. But we can (or shall, one day, be able to) make a rendering of *weather* which, though not the same as the real weather conditions prevailing on any historical day, is nevertheless so realistic in its behaviour that no user, however expert, will be able to distinguish it from genuine weather. The same is true of any environment that does not show the effects of quantum interference (which means most environments). Rendering such an environment in virtual reality is a tractable computational task. However, it would appear that no practical rendering is possible for environments that do show the effects of quantum interference. Without performing exponentially large amounts of computation, how can we be sure that in those cases our rendered environment will not do things which the real environment strictly never does, because of some interference phenomenon?

It might seem natural to conclude that reality does not, after all, display genuine computational universality, because interference phenomena cannot be usefully rendered. Feynman, however, correctly drew the opposite conclusion! Instead of regarding the intractability of the task of rendering quantum phenomena as an obstacle, Feynman regarded it as an opportunity. If it requires so much computation to work out what will happen in an interference experiment, then the very act of setting up such an experiment and measuring its outcome is tantamount to performing a complex computation. Thus, Feynman reasoned, it might after all be possible to render quantum environments efficiently, provided the computer were allowed to perform experiments on a real quantum-mechanical object. The computer would choose what measurements to make on an auxiliary piece of quantum hardware as it went along, and would incorporate the results of the measurements into its computations.

The auxiliary quantum hardware would in effect be a computer too. For example, an interferometer could act as such a device and, like any other physical object, it can be thought of as a computer. We would nowadays call it a *special-purpose quantum computer*. We 'program' it by setting up the mirrors in a certain geometry, and then projecting a single photon at the first mirror. In a non-random interference experiment the photon will always emerge in one particular direction, determined by the settings of the mirrors, and we could interpret that direction as indicating the result of the computation. In a more complex experiment, with several interacting particles, such a computation could easily, as I have explained, become 'intractable'. Yet since we could readily obtain its result just by performing this experiment, it is not really intractable after all. We must now be more careful with our terminology. Evidently there are computational tasks that are 'intractable' if we attempt to perform them using any existing computer, but which would be tractable if we were to use quantum-mechanical objects as special-purpose computers. (Notice that the fact that quantum phenomena can be used to perform computations in this way depends on their not being subject to chaos. If the outcome of computations were an inordinately sensitive function of the initial state, 'programming' the device by setting it in a suitable initial state would be an impossibly difficult task.)

Using a quantum auxiliary device in this way might be considered cheating, since *any* environment is obviously much easier to render if one has access to a spare copy of it to measure during the rendering! However, Feynman conjectured that it would not be necessary to use a literal copy of the environment being rendered: that it would be possible to find a much more easily constructed auxiliary device whose interference properties were nevertheless analogous to those of the target environment. Then a normal computer could do the rest of the rendering, working through the analogy between the auxiliary device and the target environment. And, Feynman expected, that would be a tractable task. Furthermore, he conjectured, correctly as it turned out, that all the quantum-mechanical properties of any target environment could be simulated by auxiliary devices of a particular type that he specified (namely an array of spinning atoms, each interacting with its neighbours). He called the whole class of such devices a *universal quantum simulator*.

But it was not a single machine, as it would have to be in order to qualify as a universal computer. The interactions that the simulator's atoms would have to undergo could not be fixed once and for all, as in a universal computer, but would have to be re-engineered for the simulation of each target environment. But the point of universality is that it should be possible to program a single machine, specified once and for all, to perform any possible computation, or render any physically possible environment. In 1985 I proved that under quantum physics there is a *universal quantum computer*. The proof was fairly straightforward. All I had to do was mimic Turing's constructions, but using quantum theory to define the underlying physics instead of the classical mechanics that Turing had implicitly assumed. A universal quantum computer could perform any computation that any other quantum computer (or any Turing-type computer) could perform, and it could render any finite physically possible environment in virtual reality. Moreover, it has since been shown that the time and other resources that it would need to do these things would not increase exponentially with the size or detail of the environment being rendered, so the relevant computations would be tractable by the standards of complexity theory.

The classical theory of computation, which was the unchallenged foundation of computing for half a century, is now obsolete except, like the rest of classical physics, as an approximation scheme. *The* theory of computation is now the quantum theory of computation. I said that Turing had implicitly used 'classical mechanics' in his construction. But with the benefit of hindsight we can now see that even the classical theory of computation did not fully conform to classical physics, and contained strong adumbrations of quantum theory. It is no coincidence that the word *bit*, meaning the smallest possible amount of information that a computer can manipulate, means essentially the same as *quantum*, a discrete chunk. Discrete variables (variables that cannot take a continuous range of values) are alien to classical physics. For example, if a variable has only two possible values, say 0 and 1, how does it ever get from 0 to 1? (I asked this question in Chapter 2.) In classical physics it would have to jump discontinuously, which is incompatible with how forces and motions work in classical mechanics. In quantum physics, no discontinuous change is necessary – even though all measurable quantities are discrete. It works as follows.

Let us start by imagining some parallel universes stacked like a pack of cards, with the pack as a whole representing the multiverse. (Such a model, in which the universes are arranged in a sequence, greatly understates the complexity of the multiverse, but it suffices to illustrate my point here.) Now let us alter the model to take account of the fact that the multiverse is not a discrete set of universes but a continuum, and that not all the universes are different. In fact, for each universe that is present there is also a continuum of identical universes present, comprising a certain tiny but non-zero proportion of the multiverse. In our model, this proportion may be represented by the thickness of a card, where each card now represents all the universes of a given type. However, unlike the thickness of a card, the proportion of each type of universe changes with time under quantum-mechanical laws of motion. Consequently, the proportion of universes having a given property also changes, and it changes continuously. In the case of a discrete variable changing from 0 to 1, suppose that the variable has the value 0 in all universes before the change begins, and that after the change, it has the value 1 in all universes. During the change, the proportion of universes in which the value is 0 falls smoothly from 100 per cent to zero, and the proportion in which the value is 1 rises correspondingly from zero to 100 per cent. Figure 9.4 shows a multiverse view of such a change.

It might seem from Figure 9.4 that, although the transition from 0 to 1 is objectively continuous from the multiverse perspective, it remains subjectively discontinuous from the point of view of any individual universe – as represented, say, by a horizontal line half-way up Figure 9.4. However, that is merely a limitation of the diagram, and not a real feature of what is happening. Although the diagram makes it seem that there is at each instant a particular universe that 'has just changed' from 0 to 1 because it has just 'crossed the boundary', that is not really so. It cannot be, because such a universe is strictly identical with every other universe in which the bit has value 1 at that time. So if the inhabitants of one of them were experiencing a discontinuous change, then so would the inhabitants of all the others. Therefore none of them can have such an experience. Note also that, as I shall explain in Chapter 11, the idea of anything *moving* across a diagram such as Figure 9.4, in which time is already represented, is simply a mistake. At each instant the bit has value 1 in a certain proportion of universes and 0 in another. All those universes, at all those times, are already shown in Figure 9.4. They are not moving anywhere!

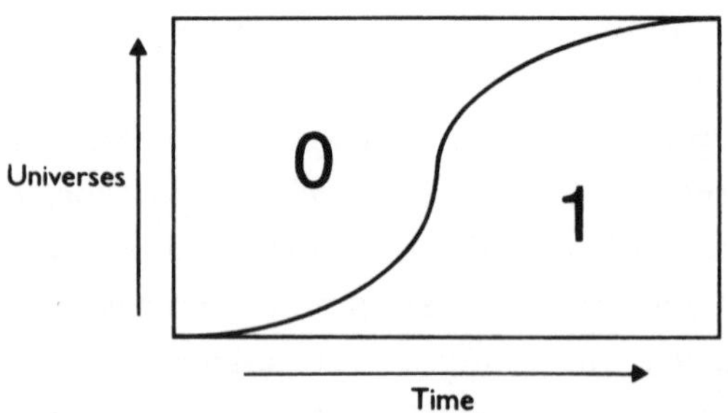

FIGURE 9.4 *Multiverse view of how a bit changes continuously from 0 to 1.*

Another way in which quantum physics is implicit in classical computation is that all practical implementations of Turing-type computers rely on such things as solid matter or magnetized materials, which could not exist in the absence of quantum-mechanical effects. For example, any solid body consists of an array of atoms, which are themselves composed of electrically charged particles (electrons, and protons in the nuclei). But because of classical chaos, no array of charged particles could be stable under classical laws of motion. The positively and negatively charged particles would simply move out of position and crash into each other, and the structure would disintegrate. It is only the strong quantum interference between the various paths taken by charged particles in parallel universes that prevents such catastrophes and makes solid matter possible.

Building a universal quantum computer is well beyond present technology. As I have said, detecting an interference phenomenon always involves setting up an appropriate interaction between *all* the variables that have been different in the universes that contribute to the interference. The more interacting particles are involved, therefore, the harder it tends to be to engineer the interaction that would display the interference – that is, the result of the computation. Among the many technical difficulties of working at the level of a single atom or single electron, one of the most important is that of preventing the environment from being affected by the different interfering sub-computations. For if a group of atoms is undergoing an interference phenomenon, and they differentially affect other atoms in the environment, then the interference can no longer be detected by measurements of the original group alone, and the group is no longer performing any useful quantum computation. This is called *decoherence*. I must add that this problem is often presented the wrong way round: we are told that 'quantum interference is a very delicate process, and must be shielded from all outside influences'. This is wrong. Outside influences could cause minor imperfections, but it is the effect of the quantum computation on the outside world that causes decoherence.

Thus the race is on to engineer sub-microscopic systems in which information-carrying variables interact among themselves, but affect their environment as little as possible. Another novel simplification, unique to the quantum theory of computation, partly offsets the difficulties caused by decoherence. It turns out that, unlike classical computation, where one needs to engineer specific classical logic elements such as AND, *or* and NOT, the precise form of the interactions hardly matters in the quantum case. Virtually any atomic-scale system of interacting bits, so long as it does not decohere, could be made to perform useful quantum computations.

Interference phenomena involving vast numbers of particles, such as superconductivity and superfluidity, are known, but it seems that none of them can be used to perform any interesting computations. At the time of writing, only single-bit quantum computations can be easily performed in the laboratory. Experimentalists are confident, however, that two- and higher-bit *quantum gates* (the quantum equivalent of the classical logical elements) will be constructed within the next few years. These are the basic components of quantum computers. Some physicists, notably Rolf Landauer of IBM Research, are pessimistic about the prospects for further advances after that. They believe that decoherence will never be reduced to the point where more than a few consecutive quantum-computational steps can be performed. Most researchers in the field are much more optimistic (though perhaps that is because only optimistic researchers choose to work on quantum computation!). Some special-purpose quantum computers have already been built (see below), and my own opinion is that more complex ones will appear in a matter of years rather than decades. As for the universal quantum computer, I expect that its construction too is only a matter of time, though I should not like to predict whether that time will be decades or centuries.

The fact that the repertoire of the universal quantum computer contains environments whose rendering is classically intractable implies that new classes of purely mathematical computations must have become tractable too. For the laws of physics are, as Galileo said, expressed in mathematical language, and rendering an environment is tantamount to evaluating certain mathematical functions. And indeed, many mathematical tasks have now been discovered which could be efficiently performed by quantum computation where all known classical methods are intractable. The most spectacular of these is the task of factorizing large numbers. The method, known as *Shor's algorithm*, was discovered in 1994 by Peter Shor of Bell Laboratories. (While this book was in proof further spectacular quantum algorithms have been discovered, including *Grover's algorithm* for searching long lists very rapidly.)

Shor's algorithm is extraordinarily simple and requires far more modest hardware than would be needed for a universal quantum computer. It is likely, therefore, that a *quantum factorization engine* will be built long before the full range of quantum computations is technologically feasible. This is a prospect of great significance for *cryptography* (the science of securely communicating and authenticating information). Realistic communication networks may be global and have large, constantly changing sets of participants with unpredictable patterns of communication. It is impractical to require every pair of participants, physically and in advance, to exchange secret cryptographic keys that would allow them later to communicate without fear of eavesdroppers. *Public-key cryptography* is any method of sending secret information where the sender and recipient do not already share any secret information. The most secure known method of public-key cryptography depends on the intractability of the problem of factorizing large numbers. This method is known as the RSA cryptosystem, named after Ronald Rivest, Adi Shamir and Leonard Adelman, who first proposed it in 1978. It depends on a mathematical procedure whereby a message can be encoded using a large (say, 250-digit) number as a key. The recipient can freely make this key public, because any message encoded with it can only be decoded given a knowledge of the factors of that number. Thus I can choose two 125-digit prime numbers and keep them secret, but multiply them together and make their 250-digit product public. Anyone can send me a message using that number as the key, but only I can read the messages because only I know the secret factors.

As I have said, there is no practical prospect of factorizing

250-digit numbers by classical means. But a quantum factorization engine running Shor's algorithm could do it using only a few thousand arithmetic operations, which might well take only a matter of minutes. So anyone with access to such a machine would easily be able to read any intercepted message that had been encrypted using the RSA cryptosystem.

It would do the cryptographers no good to choose larger numbers as keys because the resources required by Shor's algorithm increase only slowly with the size of the number being factorized. In the quantum theory of computation, factorization is a very tractable task. It is thought that, in the presence of a given level of decoherence, there would again be a practical limit on the size of number that could be factorized, but there is no known lower limit on the rate of decoherence that can be technologically achieved. So we must conclude that one day in the future, at a time that cannot now be predicted, the RSA cryptosystem with any given length of key may become insecure. In a certain sense, that makes it insecure even today. For anyone, or any organization, that records an RSA-encrypted message today, and waits until they can buy a quantum factorization engine with low enough decoherence, will be able to decode the message. That may not happen for centuries, or it may be only decades – perhaps less, who can tell? But the likelihood that it will be rather a long time is all that now remains of the former complete security of the RSA system.

When a quantum factorization engine is factorizing a 250-digit number, the number of interfering universes will be of the order of 10^{500} – that is, ten to the power of 500. This staggeringly large number is the reason why Shor's algorithm makes factorization tractable. I said that the algorithm requires only a few thousand arithmetic operations. I meant, of course, a few thousand operations *in each universe* that contributes to the answer. All those computations are performed in parallel, in different universes, and share their results through interference.

You may be wondering how we can persuade our counterparts in 10^{500}-odd universes to start working on our factorization task. Will they not have their own agendas for the use of their computers?

No – and no persuasion is necessary. Shor's algorithm operates initially only on a set of universes that are *identical* to one another, and it causes them to become differentiated only within the confines of the factorization engine. So we, who specified the number to be factorized, and who wait while the answer is computed, are identical in all the interfering universes. There are, no doubt, many other universes in which we programmed different numbers or never built the factorization engine at all. But those universes differ from ours in too many variables – or more precisely, in variables that are not made to interact in the right way by the programming of Shor's algorithm – and so do not interfere with our universe.

The argument of Chapter 2, applied to *any* interference phenomenon destroys the classical idea that there is only one universe. Logically, the possibility of complex quantum computations adds nothing to a case that is already unanswerable. But it does add psychological impact. With Shor's algorithm, the argument has been writ very large. To those who still cling to a single-universe world-view, I issue this challenge: *explain how Shor's algorithm works*. I do not merely mean predict that it will work, which is merely a matter of solving a few uncontroversial equations. I mean provide an explanation. When Shor's algorithm has factorized a number, using 10^{500} or so times the computational resources that can be seen to be present, where was the number factorized? There are only about 10^{80} atoms in the entire visible universe, an utterly minuscule number compared with 10^{500}. So if the visible universe were the extent of physical reality, physical reality would not even remotely contain the resources required to factorize such a large number. Who did factorize it, then? How, and where, was the computation performed?

I have been discussing traditional types of mathematical task that quantum computers would be able to perform more quickly than existing machines. But there is an additional class of new tasks open to quantum computers that no classical computer could perform at all. By a strange coincidence, one of the first of these tasks to be discovered also concerns public-key cryptography. This time it is not a matter of breaking an existing system, but of implementing a new, absolutely secure system of *quantum cryptography*. In 1989, at IBM Research, Yorktown Heights, New York, in the office of the theoretician Charles Bennett, the first working quantum computer was built. It was a special-purpose quantum computer consisting of a pair of quantum cryptographic devices designed by Bennett and Gilles Brassard of the University of Montreal. It became the first machine ever to perform non-trivial computations that no Turing machine could perform.

In Bennett and Brassard's quantum cryptosystem, messages are encoded in the states of individual photons emitted by a laser. Although many photons are needed to transmit a message (one photon per bit, plus many more photons wasted in various inefficiencies), the machines can be built with existing technology because they need to perform their quantum computations on only one photon at a time. The system's security is based not on intractability, either classical or quantum, but directly on the properties of quantum interference: that is what gives it its classically unobtainable absolute security. No amount of future computation by any sort of computer, whether for millions or trillions of years, would be of any help to an eavesdropper on quantum-encrypted messages: for if one communicates through a medium exhibiting interference, *one can detect eavesdroppers*. According to classical physics, there is nothing that can prevent an eavesdropper who has physical access to a communication medium, such as a telephone line, from installing a passive listening device. But, as I have explained, if one makes any measurement on a quantum system, one alters its subsequent interference properties. The communication protocol relies on this effect. The communicating parties effectively set up repeated interference experiments, co-ordinating them over a public communication channel. Only if the interference passes the test for there having been no eavesdropper do they pass on to the next stage of the protocol, which is to use some of the transmitted information as a cryptographic key. At worst, a persistent eavesdropper might prevent any communication from taking place at all (though of course that is more easily achieved just by cutting the telephone line). But as for reading a message, only the intended recipient can do that, and the guarantee of that is provided by the laws of physics.

Because quantum cryptography depends on manipulating individual photons, it suffers from a major limitation. Each photon that is successfully received, carrying one bit of the message, must somehow have been transmitted intact from the transmitter to the receiver. But every method of transmission involves losses, and if these are too heavy the message will never arrive. Setting up relay stations (which is the remedy for this problem in existing communication systems) would compromise the security because an eavesdropper could, without being detected, monitor what goes on inside the relay station. The best existing quantum-cryptographic systems use fibre-optic cables and have a range of about ten kilometres. This would suffice to provide, say, the financial district of a city with absolutely secure internal communications. Marketable systems may not be far away, but to solve the problem of public-key cryptography in general – say, for global communication – further advances in quantum cryptography are required.

Experimental and theoretical research in the field of quantum computation is accelerating world-wide. Ever more promising new technologies for realizing quantum computers are being proposed, and new types of quantum computation with various advantages over classical computation are continually being discovered and analysed. I find all these developments very exciting, and I believe that some of them will bear technological fruit. But as far as this book is concerned, that is a side-issue. From a fundamental standpoint it does not matter how useful quantum computation turns out to be, nor does it matter whether we build the first universal quantum computer next week, or centuries from now, or never. The quantum theory of computation must in any case be an integral part of the world-view of anyone who seeks a fundamental understanding of reality. What quantum computers tell us about connections between the laws of physics, universality, and apparently unrelated strands of explanation of the fabric of reality, we can discover – and are already discovering – by studying them theoretically.

TERMINOLOGY

quantum computation Computation that requires quantum-mechanical processes, especially interference. In other words, computation that is performed in collaboration between parallel universes.

exponential computation A computation whose resource requirements (such as the time required) increase by a roughly constant factor for each additional digit in the input.

tractable/intractable (Rough-and-ready rule:) A computational task is deemed tractable if the resources required to perform it do not increase exponentially with the number of digits in the input.

chaos The instability in the motion of most classical systems. A small difference between two initial states gives rise to exponentially growing deviations between the two resulting trajectories. But reality obeys quantum and not classical physics. Unpredictability caused by chaos is in general swamped by quantum indeterminacy caused by identical universes becoming different.

universal quantum computer A computer that could perform any computation that any other quantum computer could perform, and render any finite, physically possible environment in virtual reality.

quantum cryptography Any form of cryptography that can be performed by quantum computers but not by classical computers.

special-purpose quantum computer A quantum computer, such as a quantum cryptographic device or quantum factorization engine, that is not a universal quantum computer.

decoherence If different branches of a quantum computation, in different universes, affect the environment differently, then interference is reduced and the computation may fail. Decoherence is the principal obstacle to the practical realization of more powerful quantum computers.

SUMMARY

The laws of physics permit computers that can render every physically possible environment without using impractically large resources. So universal computation is not merely possible, as required by the Turing principle, it is also tractable. Quantum phenomena may involve vast numbers of parallel universes and therefore may not be capable of being efficiently simulated within one universe. However, this strong form of universality still holds because quantum computers can efficiently render every physically possible quantum environment, even when vast numbers of universes are interacting. Quantum computers can also efficiently solve certain mathematical problems, such as factorization, which are classically intractable, and can implement types of cryptography which are classically impossible. Quantum computation is a qualitatively new way of harnessing nature.

The next chapter is likely to provoke many mathematicians. This can't be helped. Mathematics is not what they think it is.

(Readers who are unfamiliar with traditional assumptions about the certainty of mathematical knowledge may consider the chapter's main conclusion – that our knowledge of mathematical truth depends on, and is no more reliable than, our knowledge of the physical world – to be obvious. Such readers may prefer to skim this chapter and hurry on to the discussion of time in Chapter 11.)

10

The Nature of Mathematics

The 'fabric of reality' that I have been describing so far has been the fabric of *physical* reality. Yet I have also referred freely to entities that are nowhere to be found in the physical world – abstractions such as numbers and infinite sets of computer programs. Nor are the laws of physics themselves physical entities in the sense that rocks and planets are. As I have said, Galileo's 'Book of Nature' is only a metaphor. And then there are the fictions of virtual reality, the non-existent environments whose laws differ from the real laws of physics. Beyond those are what I have called the 'Cantgotu' environments, which cannot even be rendered in virtual reality. I have said that there exist infinitely many of those for every environment that can be rendered. But what does it mean to say that such environments 'exist'? If they do not exist in reality, or even in virtual reality, where do they exist?

Do abstract, non-physical entities exist? Are they part of the fabric of reality? I am not interested here in issues of mere word usage. It is obvious that numbers, the laws of physics, and so on do 'exist' in some senses and not in others. The substantive question is this: how are we to understand such entities? Which of them are merely convenient forms of words, referring ultimately only to ordinary, physical reality? Which are merely ephemeral features of our culture? Which are arbitrary, like the rules of a trivial game that we need only look up? And which, if any, can be explained only in a way that attributes an independent existence to them? Things of this last type *must* be part of the fabric of reality as defined in this book, because one would have to understand them in order to understand everything that is understood.

This suggests that we ought to apply Dr Johnson's criterion again. If we want to know whether a given abstraction really exists, we should ask whether it 'kicks back' in a complex, autonomous way. For example, mathematicians characterize the 'natural numbers' 1, 2, 3, . . . in the first instance through a precise definition such as:

> 1 is a natural number.
>
> Each natural number has precisely one successor, which is also a natural number.
>
> 1 is not the successor of any natural number.
>
> Two natural numbers with the same successor are the same.

Such definitions are attempts to express abstractly the intuitive *physical* notion of successive amounts of a discrete quantity. (More precisely, as I explained in the previous chapter, that notion is really quantum-mechanical.) The operations of arithmetic, such as multiplication and addition, and further concepts such as that of a prime number, are then defined with reference to the 'natural numbers'. But having created abstract 'natural numbers' through that definition, and having understood them through that intuition, we find that there is a lot more that we still do not understand about them. The definition of a prime number fixes once and for all which numbers are primes and which are not. But the *understanding* of which numbers are prime – for instance, how prime numbers are distributed on very large scales, how clumped they are, how 'random' they are, and why – involves a wealth of new insights and new explanations. Indeed, it turns out that number theory is a whole world (the term is often used) in itself. To understand numbers more fully we have to define many new classes of abstract entities, and postulate many new structures and connections among those structures. We find that some of these abstract structures are related to other intuitions that we already had but which, on the face of it, had nothing to do with numbers – such as *symmetry*, *rotation*, the *continuum*, *sets*, *infinity*, and many more. Thus, abstract mathematical entities we think we are familiar with can nevertheless surprise or disappoint us. They can pop up unexpectedly in new guises, or disguises. They can be inexplicable, and then later conform to a new explanation. So they are complex and autonomous, and therefore by Dr Johnson's criterion we must conclude that they are real. Since we cannot understand them either as being part of ourselves or as being part of something else that we already understand, but we *can* understand them as independent entities, we must conclude that they *are* real, independent entities.

Nevertheless, abstract entities are intangible. They do not kick back physically in the sense that a stone does, so experiment and observation cannot play quite the same role in mathematics as they do in science. In mathematics, *proof* plays that role. Dr Johnson's stone kicked back by making his foot rebound. Prime numbers kick back when we prove something unexpected about them – especially if we can go on to explain it too. In the traditional view, the crucial difference between proof and experiment is that a proof makes no reference to the physical world. We can perform a proof in the privacy of our own minds, or we can perform a proof trapped inside a virtual-reality generator rendering the wrong physics. Provided only that we follow the rules of mathematical inference, we should come up with the same answer as anyone else. And again, the prevailing view is that, apart from the possibility of making blunders, when we have proved something we know with *absolute certainty* that it is true.

Mathematicians are rather proud of this absolute certainty, and scientists tend to be a little envious of it. For in science there is no way of being certain of any proposition. However well one's theories explain existing observations, at any moment someone may make a new, inexplicable observation that casts doubt on the whole of the current explanatory structure. Worse, someone may reach a better understanding that explains not only all existing observations but also why the previous explanations seemed to work but are nevertheless quite wrong. Galileo, for instance, found a new explanation of the age-old observation that the ground beneath our feet is at rest, an explanation that involved the ground actually moving. Virtual reality – which can make one environment seem to be another – underlines the fact that when observation is the ultimate arbiter between theories, there can never be any certainty that an existing explanation, however obvious, is even remotely true. But when proof is the arbiter, it is supposed, there can be certainty.

It is said that the rules of logic were first formulated in the hope that they would provide an impartial and infallible method of resolving all disputes. This hope can never be fulfilled. The study of logic itself revealed that the scope of logical deduction as a means of discovering the truth is severely limited. Given substantive assumptions about the world, one can deduce conclusions; but the conclusions are no more secure than the assumptions. The only propositions that logic can prove without recourse to assumptions are tautologies – statements such as 'all planets are planets', which assert nothing. In particular, all substantive questions of science lie outside the domain in which logic alone can settle disputes. But mathematics, it is supposed, lies *inside* that domain. Thus mathematicians seek absolute but abstract truth, while scientists console themselves with the thought that they can gain substantive and useful knowledge of the physical world. But they must accept that this knowledge comes without guarantees. It is forever tentative, forever fallible. The idea that science is characterized by 'induction', a method of justification which is supposed to be a slightly fallible analogue of logical deduction, is an attempt to make the best of this perceived second-class status of scientific knowledge. Instead of deductively justified certainties, perhaps we can make do with inductively justified near-certainties.

As I have said, there is no such method of justification as 'induction'. The idea of reasoning one's way to 'near-certainty' in science is myth. How could I prove with 'near-certainty' that a wonderful new theory of physics, overturning my most unquestioned assumptions about reality, will not be published tomorrow? Or that I am not inside a virtual-reality generator? But all this is not to say that scientific knowledge is indeed 'second-class'. For the idea that mathematics yields certainties *is a myth too*.

Since ancient times, the idea that mathematical knowledge has a privileged status has often been associated with the idea that some abstract entities, at least, are not merely part of the fabric of reality but are even more real than the physical world. Pythagoras believed that regularities in nature are the expression of mathematical relationships between natural numbers. 'All things are numbers' was the slogan. This was not meant quite literally, but Plato went further and effectively denied that the physical world is real at all. He regarded our apparent experiences of it as worthless or misleading, and argued that the physical objects and phenomena we perceive are merely 'shadows' or imperfect imitations of their ideal essences ('Forms', or 'Ideas') which exist in a separate realm that is the true reality. In that realm there exist, among other things, the Forms of pure numbers such as 1, 2, 3, . . . , and the Forms of mathematical operations such as addition and multiplication. We may perceive some shadows of these Forms, as when we place an apple on the table, and then another apple, and then see that there are two apples. But the apples exhibit 'one-ness' and 'two-ness' (and, for that matter, 'apple-ness') only imperfectly. They are not perfectly identical, so there are never really *two* of anything on the table. It might be objected that the number two could also be represented by there being two *different* things on the table. But such a representation is still imperfect because we must then admit that there are cells that have fallen from the apples, and dust, and air, on the table as well. Unlike Pythagoras, Plato had no particular axe to grind about the natural numbers. His reality contained the Forms of all concepts. For example, it contained the Form of a perfect circle. The 'circles' we experience are never really circles. They are not perfectly round, nor perfectly planar; they have a finite thickness; and so on. All of them are imperfect.

Plato then pointed out a problem. Given all this Earthly imperfection (and, he could have added, given our imperfect sensory access even to Earthly circles), how can we possibly know what we know about real, perfect circles? Evidently we do know about them, but how? Where did Euclid obtain the knowledge of geometry which he expressed in his famous axioms, when no genuine circles, points or straight lines were available to him? Where does the certainty of a mathematical proof come from, if no one can perceive the abstract entities that the proof refers to? Plato's answer was that we do not obtain our knowledge of such things from this world of shadow and illusion. Instead, we obtain it directly from the real world of Forms itself. We have perfect inborn knowledge of that world which is, he suggests, forgotten at birth, and then obscured by layers of errors caused by trusting our senses. But reality can be remembered through the diligent application of 'reason', which then yields the absolute certainty that experience can never provide.

I wonder whether anyone has ever believed this rather rickety fantasy (including Plato himself, who was after all a very competent philosopher who believed in telling ennobling lies to the public). However, the problem he posed – of how we can possibly have knowledge, let alone certainty, of abstract entities – is real enough, and some elements of his proposed solution have been part of the prevailing theory of knowledge ever since. In particular, the core idea that mathematical knowledge and scientific knowledge come from *different sources*, and that the 'special' source of mathematics confers *absolute certainty* upon it, is to this day accepted uncritically by virtually all mathematicians. Nowadays they call this source *mathematical intuition*, but it plays exactly the same role as Plato's 'memories' of the realm of Forms.

There have been many bitter controversies about precisely which types of perfectly reliable knowledge our mathematical intuition can be expected to reveal. In other words, mathematicians agree that mathematical intuition is a source of absolute certainty, but they cannot agree about what mathematical intuition tells them! Obviously this is a recipe for infinite, unresolvable controversy.

Inevitably, most such controversies have centred on the validity or otherwise of various methods of proof. One controversy concerned so-called 'imaginary' numbers. Imaginary numbers are the square roots of negative numbers. New theorems about ordinary, 'real' numbers were proved by appealing, at intermediate stages of a proof, to the properties of imaginary numbers. For example, the first theorems about the distribution of prime numbers were proved in this way. But some mathematicians objected to imaginary numbers on the grounds that they were not real. (Current terminology still reflects the old controversy, even though we now think that imaginary numbers are just as real as 'real' numbers.) I expect that their schoolteachers had told them that they were not *allowed* to take the square root of minus one, and consequently they did not see why anyone else should be allowed to. No doubt they called this uncharitable impulse 'mathematical intuition'. But other mathematicians had different intuitions. They understood what the imaginary numbers were and how they fitted in with the real numbers. Why, they thought, should one not define new abstract entities to have any properties one likes? Surely the only legitimate grounds for forbidding this would be that the required properties were logically inconsistent. (That is essentially the modern consensus which the mathematician John Horton Conway has robustly referred to as the 'Mathematicians' Liberation Movement'.) Admittedly, no one had proved that the system of imaginary numbers *was* self-consistent. But then, no one had *proved* that the ordinary arithmetic of the natural numbers was self-consistent either.

There were similar controversies over the validity of the use of infinite numbers, and of sets containing infinitely many elements, and of the infinitesimal quantities that were used in calculus. David Hilbert, the great German mathematician who provided much of the mathematical infrastructure of both the general theory of relativity and quantum theory, remarked that 'the literature of mathematics is glutted with inanities and absurdities which have had their source in the infinite'. Some mathematicians, as we shall see, denied the validity of reasoning about infinite entities at all. The runaway success of pure mathematics during the nineteenth century had done little to resolve these controversies. On the contrary, it tended to intensify them and raise new ones. As mathematical reasoning became more sophisticated, it inevitably moved ever further away from everyday intuition, and this had two important, opposing effects. First, mathematicians became more meticulous about proofs, which were subjected to ever increasing standards of rigour before they were accepted. But second, more powerful *methods* of proof were invented which could not always be validated by existing methods. And that often raised doubts as to whether a particular method of proof, however self-evident, was completely infallible.

So by about 1900 there was a crisis at the foundations of mathematics – namely, that there were no foundations. But what had become of the laws of pure logic? Were they not supposed to resolve all disputes within the realm of mathematics? The embarrassing fact was that the 'laws of pure logic' were in effect what the disputes in mathematics were now about. Aristotle had been the first to codify such laws in the fourth century BC, and so founded what is today called *proof theory*. He assumed that a proof must consist of a sequence of statements, starting with some premises and definitions and ending with the desired conclusion. For a sequence of statements to be a valid proof, each statement, apart from the premises at the beginning, had to follow from previous ones according to one of a fixed set of patterns called *syllogisms*. A typical syllogism was

All men are mortal.
Socrates is a man.
———
[Therefore] *Socrates is mortal.*

In other words, this rule said that if a statement of the form 'all As have property B' (as in 'all men are mortal') appears in a proof, and another statement of the form 'the individual X is an A' (as in 'Socrates is a man') also appears, then the statement 'X has property B' ('Socrates is mortal') may validly appear later in the proof, and in particular it is a valid conclusion. The syllogisms expressed what we would call *rules of inference* – that is, rules defining the steps that are permitted in proofs, such that the truth of the premises is transmitted to the conclusions. By the same token, they are rules that can be applied to determine whether a purported proof is valid or not.

Aristotle had declared that all valid proofs could be expressed in syllogistic form. But he had not proved this! And the problem for proof theory was that very few modern mathematical proofs were expressed purely as a sequence of syllogisms; nor could many of them be recast in that form, even in principle. Yet most mathematicians could not bring themselves to stick to the letter of the Aristotelian law, since some of the new proofs seemed just as self-evidently valid as Aristotelian reasoning. Mathematics had moved on. New tools such as symbolic logic and set theory allowed mathematicians to relate mathematical structures to one another in new ways. This had created new self-evident truths that were independent of the classical rules of inference, so those classical rules were self-evidently inadequate. But which of the new methods of proof were genuinely infallible? How were the rules of inference to be modified so that they would have the completeness that Aristotle had mistakenly claimed? How could the absolute authority of the old rules ever be regained if mathematicians could not agree on what was self-evident and what was nonsense?

Meanwhile, mathematicians were continuing to construct their abstract castles in the sky. For practical purposes many of these constructs seemed sound enough. Some had become indispensable in science and technology, and most were connected by a beautiful and fruitful explanatory structure. Nevertheless, no one could guarantee that the entire structure, or any substantial part of it, was not founded upon a logical contradiction, which would make it literally nonsense. In 1902 Bertrand Russell proved that a scheme for defining set theory rigorously, which had just been proposed by the German logician Gottlob Frege, was inconsistent. This did not mean that it was necessarily invalid to use sets in proofs. Indeed, very few mathematicians seriously supposed that any of the usual ways of using sets, or arithmetic, or other core areas of mathematics, might be invalid. What was shocking about Russell's result was that mathematicians had believed their subject to be *par excellence* the means of delivering absolute certainty through the proofs of mathematical theorems. The very possibility of controversy over the validity of different methods of proof undermined the whole purpose (as it was supposed) of the subject.

Many mathematicians therefore felt that it was a matter of urgency to place proof theory, and thereby mathematics itself, on a secure foundation. They wanted to consolidate after their headlong advances: to define once and for all which types of proof were absolutely secure, and which were not. Whatever was outside the secure zone could be dropped, and whatever was inside would be the sole basis of all future mathematics.

To this end, the Dutch mathematician Luitzen Egbertus Jan Brouwer advocated an extreme conservative strategy for proof theory, known as *intuitionism*, which still has adherents to this day. Intuitionists try to construe 'intuition' in the narrowest conceivable way, retaining only what they consider to be its unchallengeably self-evident aspects. Then they elevate mathematical intuition, thus defined, to a status higher even than Plato afforded it: they regard it as being prior even to pure logic. Thus they regard logic itself as untrustworthy, except where it is justified by direct mathematical intuition. For instance, intuitionists deny that it is possible to have a direct intuition of any infinite entity. Therefore they deny that any infinite sets, such as the set of all natural numbers, exist at all. The proposition 'there exist infinitely many natural numbers' they would consider self-evidently false. And the proposition 'there exist more Cantgotu environments than physically possible environments' they would consider completely meaningless.

Historically, intuitionism played a valuable liberating role, just as inductivism did. It dared to question received certainties – some of which were indeed false. But as a positive theory of what is or is not a valid mathematical proof, it is worthless. Indeed, intuitionism is precisely the expression, in mathematics, of solipsism. In both cases there is an over-reaction to the thought that we cannot be *sure* of what we know about the wider world. In both cases the proposed solution is to retreat into an inner world which we can supposedly know directly and therefore (?) can be sure of knowing truly. In both cases the solution involves either denying the existence – or at least renouncing explanation – of what lies outside. And in both cases this renunciation also makes it impossible to explain much of what lies inside the favoured domain. For instance, if it is indeed false, as intuitionists maintain, that there exist infinitely many natural numbers, then we can infer that there must be only finitely many of them. How many? And then, however many there are, why can we not form an intuition of the next natural number above that one? Intuitionists would explain this problem away by pointing out that the argument I have just given assumes the validity of ordinary logic. In particular, it involves inferring, from the fact that there are not infinitely many natural numbers, that there must be some particular finite number of them. The relevant rule of inference is called the *law of the excluded middle*. It says that, for any proposition X (such as 'there are infinitely many natural numbers'), there is no third possibility between X being true and its negation ('there are finitely many natural numbers') being true. Intuitionists coolly deny the law of the excluded middle.

Since, in most people's minds, the law of the excluded middle is itself backed by a powerful intuition, its rejection naturally causes non-intuitionists to wonder whether the intuitionists' intuition is so self-evidently reliable after all. Or, if we consider the law of the excluded middle to stem from a *logical* intuition, it leads us to re-examine the question whether mathematical intuition really supersedes logic. At any rate, can it be *self-evident* that it does?

But all that is only to criticize intuitionism from the outside. It is no disproof; nor can intuitionism ever be disproved. If someone insists that a self-consistent proposition is self-evident to them, just as if they insist that they alone exist, they cannot be proved wrong. However, as with solipsism generally, the truly fatal flaw of intuitionism is revealed not when it is attacked, but when it is taken seriously in its own terms, as an explanation of its own, arbitrarily truncated world. Intuitionists believe in the reality of the finite natural numbers 1, 2, 3, . . . , and even 10,949,769,651,859. But the intuitive argument that because each of these numbers has a successor, they form an infinite sequence, is in the intuitionists' view no more than a self-delusion or affectation and is literally untenable. But by severing the link between their version of the abstract 'natural numbers' and the intuitions that those numbers were originally intended to formalize, intuitionists have also denied themselves the usual explanatory structure through which natural numbers are understood. This raises a problem for anyone who prefers explanations to unexplained complications. Instead of solving that problem by providing an alternative or deeper explanatory structure for the natural numbers, intuitionism does exactly what the Inquisition did, and what solipsists do: it retreats still further from explanation. It introduces further unexplained complications (in this case the denial of the law of the excluded middle) whose only purpose is to allow intuitionists to behave as if their opponents' explanation were true, while drawing no conclusions about reality from this.

Just as solipsism starts with the motivation of simplifying a frighteningly diverse and uncertain world, but when taken seriously turns out to be realism *plus* some unnecessary complications, so intuitionism ends up being one of the most counter-intuitive doctrines that has ever been seriously advocated.

David Hilbert proposed a much more commonsensical – but still ultimately doomed – plan to 'establish once and for all the certitude of mathematical methods'. Hilbert's plan was based on the idea of consistency. He hoped to lay down, once and for all, a complete set of modern rules of inference for mathematical proofs, with certain properties. They would be finite in number. They would be straightforwardly applicable, so that determining whether any purported proof satisfied them or not would be an uncontroversial exercise. Preferably, the rules would be intuitively self-evident, but that was not an overriding consideration for the pragmatic Hilbert. He would be satisfied if the rules corresponded only moderately well to intuition, provided that he could be sure that they were self-consistent. That is, if the rules designated a given proof as valid, he wanted to be sure that they could never designate any proof with the opposite conclusion as valid. How could he be sure of such a thing? This time, consistency would have to be *proved*, using a method of proof which itself adhered to the same rules of inference. Then Hilbert hoped that Aristotelian completeness and certainty would be restored, and that every true mathematical statement would in principle be provable under the rules, and that no false statement would be. In 1900, to mark the turn of the century, Hilbert published a list of problems that he hoped mathematicians might be able to solve during the course of the twentieth century. The tenth problem was to find a set of rules of inference with the above properties, and, by their own standards, to prove them consistent.

Hilbert was to be definitively disappointed. Thirty-one years later, Kurt Gödel revolutionized proof theory with a root-and-branch refutation from which the mathematical and philosophical worlds are still reeling: he proved that Hilbert's tenth problem is insoluble. Gödel proved first that any set of rules of inference that is capable of correctly validating even the proofs of ordinary arithmetic could never validate a proof of its own consistency. Therefore there is no hope of finding the provably consistent set of rules that Hilbert envisaged. Second, Gödel proved that if a set of rules of inference in some (sufficiently rich) branch of mathematics *is* consistent (whether provably so or not), then within that branch of mathematics there must exist valid methods of proof that those rules fail to designate as valid. This is called *Gödel's incompleteness theorem*. To prove his theorems, Gödel used a remarkable extension of the Cantor 'diagonal argument' that I mentioned in Chapter 6. He began by considering any consistent set of rules of inference. Then he showed how to construct a proposition which could neither be proved nor disproved under those rules. Then he proved that that proposition would be true.

If Hilbert's programme had worked, it would have been bad news for the conception of reality that I am promoting in this book, for it would have removed the necessity for *understanding* in judging mathematical ideas. Anyone – or any mindless machine – that could learn Hilbert's hoped-for rules of inference by heart would be as good a judge of mathematical propositions as the ablest mathematician, yet without needing the mathematician's insight or understanding, or even having the remotest clue as to what the propositions were about. In principle, it would be possible to make new mathematical discoveries without knowing any mathematics at all, beyond Hilbert's rules. One would simply check through all possible strings of letters and mathematical symbols in alphabetical order, until one of them passed the test for being a proof or disproof of some famous unsolved conjecture. In principle, one could settle any mathematical controversy without ever understanding it – without even knowing the meanings of the symbols, let alone understanding how the proof worked, or what it proved, or what the method of proof was, or why it was reliable.

It may seem that the achievement of a unified standard of proof in mathematics could at least have helped us in the overall drive towards unification – that is, the 'deepening' of our knowledge that I referred to in Chapter 1. But the opposite is the case. Like the predictive 'theory of everything' in physics, Hilbert's rules would have told us almost nothing about the fabric of reality. They would, as far as mathematics goes, have realized the ultimate reductionist vision, predicting everything (in principle) but explaining nothing. Moreover, if mathematics had been reductionist then all the undesirable features which I argued in Chapter 1 are absent from the structure of human knowledge would have been present in mathematics: mathematical ideas would have formed a hierarchy, with Hilbert's rules at its root. Mathematical truths whose verification from the rules was very complex would have been objectively less fundamental than those that could be verified immediately from the rules. Since there could have been only a finite supply of such fundamental truths, as time went on mathematics would have had to concern itself with ever less fundamental problems. Mathematics might well have come to an end, under this dismal hypothesis. If it did not, it would inevitably have fragmented into ever more arcane specialities, as the complexity of the 'emergent' issues that mathematicians would have been forced to study increased, and as the connections between those issues and the foundations of the subject became ever more remote.

Thanks to Gödel, we know that there will never be a fixed method of determining whether a mathematical proposition is true, any more than there is a fixed way of determining whether a scientific theory is true. Nor will there ever be a fixed way of generating new mathematical knowledge. Therefore progress in mathematics will always depend on the exercise of creativity. It will always be possible, and necessary, for mathematicians to invent new types of proof. They will validate them by new arguments and by new modes of explanation depending on their ever improving understanding of the abstract entities involved. Gödel's own theorems were a case in point: to prove them, he had to invent a new method of proof. I said the method was based on the 'diagonal argument', but Gödel extended that argument in a new way. Nothing had ever been proved in this way before; no rules of inference laid down by someone who had never seen Gödel's method could possibly have been prescient enough to designate it as valid. Yet it *is* self-evidently valid. Where did this self-evidentness come from? It came from Gödel's understanding of the nature of proof. Gödel's proofs are as compelling as any in mathematics, but only if one first understands the explanation that accompanies them.

So explanation does, after all, play the same paramount role in pure mathematics as it does in science. Explaining and understanding the world – the physical world and the world of mathematical abstractions – is in both cases the object of the exercise. Proof and observation are merely means by which we check our explanations.

Roger Penrose has drawn a further, radical and very Platonic lesson from Gödel's results. Like Plato, Penrose is fascinated by the ability of the human mind to grasp the abstract certainties of mathematics. Unlike Plato, Penrose does not believe in the supernatural, and takes it for granted that the brain is part of, and has access only to, the natural world. So the problem is even more acute for him than it was for Plato: how can the fuzzy, unreliable physical world deliver mathematical certainties to a fuzzy, unreliable part of itself such as a mathematician? In particular, Penrose wonders how we can possibly perceive the infallibility of new, valid *forms* of proof, of which Gödel assures us there is an unlimited supply.

Penrose is still working on a detailed answer, but he does claim that the very existence of this sort of open-ended mathematical intuition is fundamentally incompatible with the existing structure of physics, and in particular that it is incompatible with the Turing principle. His argument, in summary, runs as follows. If the Turing principle is true, then we can consider the brain (like any other object) to be a computer executing a particular program. The brain's interactions with the environment constitute the inputs and outputs of the program. Now consider a mathematician in the act of deciding whether some newly proposed type of proof is valid or not. Making such a decision is tantamount to executing a proof-validating computer program within the mathematician's brain. Such a program embodies a set of Hilbertian rules of inference which, according to Gödel's theorem, cannot possibly be complete. Moreover, as I have said, Gödel provides a way of constructing, and proving, a true proposition which those rules can never recognize as proven. Therefore the mathematician, whose mind is effectively a computer applying those rules, can never recognize the proposition as proven either. Penrose then proposes to show the proposition, and Gödel's method of proving it to be true, to that very mathematician. The mathematician understands the proof. It is, after all, self-evidently valid, so the mathematician can presumably see that it is valid. But that would contradict Gödel's theorem. Therefore there must be a false assumption somewhere in the argument, and Penrose thinks that the false assumption is the Turing principle.

Most computer scientists do not agree with Penrose that the Turing principle is the weakest link in his story. They would say that the mathematician in the story would indeed be unable to recognize the Gödelian proposition as proven. It may seem odd that a mathematician should suddenly become unable to comprehend a self-evident proof. But look at this proposition:

David Deutsch cannot consistently judge this statement to be true.

I am trying as hard as I can, but I cannot consistently judge it to be true. For if I did, I would be judging that I *cannot* judge it to be true, and would be contradicting myself. But *you* can see that it is true, can't you? This shows it is at least possible for a proposition to be unfathomable to one person yet self-evidently true to everyone else.

Anyway, Penrose hopes for a new, fundamental theory of physics replacing both quantum theory and the general theory of relativity. It would make new, testable predictions, though it would of course agree with quantum theory and relativity for all existing observations. (There are no known experimental counter-examples to those theories.) However, Penrose's world is fundamentally very different from what existing physics describes. Its basic fabric of reality is what *we* call the world of mathematical abstractions. In this respect Penrose, whose reality includes all mathematical abstractions, but perhaps not *all* abstractions (like honour and justice), is somewhere between Plato and Pythagoras. What we call the physical world is, to him, fully real (another difference from Plato), but is somehow part of, or emergent from, mathematics itself. Furthermore, there is no universality; in particular, there is no machine that can render all possible human thought processes. Nevertheless, the world (especially, of course, its mathematical substrate) is still comprehensible. Its comprehensibility is ensured not by the universality of computation, but by a phenomenon quite new to physics (though not to Plato): *mathematical entities impinge directly on the human brain*, via physical processes yet to be discovered. In this way the brain, according to Penrose, does not do mathematics solely by reference to what we currently call the physical world. It has direct access to a Platonic reality of mathematical Forms, and can perceive mathematical truths there with (blunders aside) absolute certainty.

It is often suggested that the brain may be a quantum computer, and that its intuitions, consciousness and problem-solving abilities might depend on quantum computations. This *could* be so, but I know of no evidence and no convincing argument that it is so. My bet is that the brain, considered as a computer, is a classical one. But that issue is independent of Penrose's ideas. He is not arguing that the brain is a new sort of universal computer, differing from the universal quantum computer by having a larger repertoire of computations made possible by new, post-quantum physics. He is arguing for a new physics that will not support computational universality, so that under his new theory it will not be possible to construe some of the actions of the brain as computations at all.

I must admit that I cannot conceive of such a theory. However, fundamental breakthroughs do tend to be hard to conceive of before they occur. Naturally, it is hard to judge Penrose's theory before he succeeds in formulating it fully. If a theory with the properties he hopes for does eventually supersede quantum theory or general relativity, or both, whether through experimental testing or by providing a deeper level of explanation, then every reasonable person would want to adopt it. And then we would embark on the adventure of comprehending the new world-view that the theory's explanatory structures would compel us to adopt. It is likely that this would be a very different world-view from the one I am presenting in this book. However, even if all this came to pass, I am nevertheless at a loss to see how the theory's original motivation, that of explaining our ability to grasp new mathematical proofs, could possibly be satisfied. The fact would remain that, now and throughout history, great mathematicians have had different, conflicting intuitions about the validity of various methods of proof. So even if it is true that an absolute, physico-mathematical reality feeds its truths directly into our brains to create mathematical intuitions, mathematicians are not always able to distinguish those intuitions from other, mistaken intuitions and ideas. There is, unfortunately, no bell that rings, or light that flashes, when we are comprehending a truly valid proof. We might sometimes feel such a flash, at a 'eureka' moment – and nevertheless be mistaken. And even if the theory predicted that there *is* some previously unnoticed physical indicator accompanying true intuitions (this is getting extremely implausible now), we should certainly find it useful, but that would still not amount to a proof that the indicator works. Nothing could prove that an even better physical theory would not one day supersede Penrose's, and reveal that the supposed indicator was unreliable after all, and some other indicator was better. Thus, even if we make every possible concession to Penrose's proposal, if we imagine it is true and view the world entirely in its terms, it still does not help us to explain the alleged certainty of the knowledge that we acquire by doing mathematics.

I have presented only a sketch of the arguments of Penrose and his opponents. The reader will have gathered that essentially I side with the opponents. However, even if it is conceded that Penrose's Gödelian argument fails to prove what it sets out to prove, and his proposed new physical theory seems unlikely to explain what it sets out to explain, Penrose is nevertheless right that any world-view based on the existing conception of scientific rationality creates a problem for the accepted foundations of mathematics (or, as Penrose would have it, vice versa). This is the ancient problem that Plato raised, a problem which, as Penrose points out, becomes more acute in the light of both Gödel's theorem and the Turing principle. It is this: in a reality composed of physics and understood by the methods of science, where does mathematical certainty come from? While most mathematicians and computer scientists take the certainty of mathematical intuition for granted, they do not take seriously the problem of reconciling this with a scientific world-view. Penrose does take it seriously, and he proposes a solution. His proposal envisages a comprehensible world, rejects the supernatural, recognizes creativity as being central to mathematics, ascribes objective reality both to the physical world and to abstract entities, and involves an integration of the foundations of mathematics and physics. In all those respects I am on his side.

Since Brouwer's, and Hilbert's, and Penrose's and all other attempts to meet Plato's challenge do not seem to have succeeded, it is worth looking again at Plato's apparent demolition of the idea that mathematical truth can be obtained by the methods of science.

First of all, Plato tells us that since we have access only to imperfect circles (say) we cannot thereby obtain any knowledge of perfect circles. But why not, exactly? One might as well say that we cannot discover the laws of planetary motion because we do not have access to real planets but only to images of planets. (The Inquisition *did* say this, and I have explained why they were wrong.) One might as well say that it is impossible to build accurate machine tools because the first one would have to be built with inaccurate machine tools. With the benefit of hindsight, we can see that this line of criticism depends on a very crude picture of how science works – something like inductivism – which is hardly surprising, since Plato lived before anything that we would recognize as science. If, say, the only way of learning about circles from experience were to examine thousands of physical circles and then, from the accumulated data, to try to infer something about their abstract Euclidean counterparts, Plato would have a point. But if we form a hypothesis that real circles resemble the abstract ones in specified ways, and we happen to be right, then we may well learn something about abstract circles by looking at real ones. In Euclidean geometry one often uses diagrams to specify a geometrical problem or its solution. There is a possibility of error in such a method of description if the imperfections of circles in the diagram give a misleading impression – for example if two circles seem to touch each other when they do not. But if one understands the relationship between real circles and perfect circles, one can, with care, eliminate all such errors. If one does not understand that relationship, it is practically impossible to understand Euclidean geometry at all.

The reliability of the knowledge of a *perfect* circle that one can gain from a *diagram* of a circle depends entirely on the accuracy of the hypothesis that the two resemble each other in the relevant ways. Such a hypothesis, referring to a physical object (the diagram), amounts to a physical theory and can never be known with certainty. But that does not, as Plato would have it, preclude the possibility of learning about perfect circles from experience; it just precludes the possibility of certainty. That should not worry anyone who is looking not for certainty but for explanations.

Euclidean geometry can be abstractly formulated entirely without diagrams. But the way in which numerals, letters and mathematical symbols are used in a symbolic proof can generate no more certainty than a diagram can, and for the same reason. The symbols too are physical objects – patterns of ink on paper, say – which denote abstract objects. And again, we are relying entirely upon the hypothesis that the physical behaviour of the symbols corresponds to the behaviour of the abstractions they denote. Therefore the reliability of what we learn by manipulating those symbols depends entirely on the accuracy of our theories of their physical behaviour, and of the behaviour of our hands, eyes, and so on with which we manipulate and observe the symbols. Trick ink that caused the occasional symbol to change its appearance when we were not looking – perhaps under the remote control of some high-technology practical joker – could soon mislead us about what we know 'for certain'.

Now let us re-examine another assumption of Plato's: the assumption that we do not have access to perfection in the physical world. He may be right that we shall not find perfect honour or justice, and he is certainly right that we shall not find the laws of physics or the set of all natural numbers. But we can find a perfect hand in bridge, or the perfect move in a given chess position. That is to say, we can find physical objects or processes that fully possess the properties of the specified abstractions. We can learn chess just as well with a real chess set as we could with a perfect Form of a chess set. The fact that a knight is chipped does not make the checkmate it delivers any less final.

As it happens, a perfect Euclidean circle *can* be made available to our senses. Plato did not realize this because he did not know about virtual reality. It would not be especially difficult to program the virtual-reality generators I envisaged in Chapter 5 with the rules of Euclidean geometry in such a way that the user could experience an interaction with a perfect circle. Having no thickness, the circle would be invisible unless we also modified the laws of optics, in which case we might give it a glow to let the user know where it is. (Purists might prefer to manage without this embellishment.) We could make the circle rigid and impenetrable, and the user could test its properties using rigid, impenetrable tools and measuring instruments. Virtual-reality callipers would have to come to a perfect knife-edge so that they could measure a zero thickness accurately. The user could be allowed to 'draw' further circles or other geometrical figures according to the rules of Euclidean geometry.

The sizes of the tools, and the user's own size, could be adjustable at will, to allow the predictions of geometrical theorems to be checked on any scale, no matter how fine. In every way, the rendered circle could respond precisely as specified in Euclid's axioms. So, on the basis of present-day science we must conclude that Plato had it backwards. We *can* perceive perfect circles in physical reality (i.e. virtual reality); but we shall never perceive them in the domain of Forms, for, in so far as such a domain can be said to exist, we have no perceptions of it at all.

Incidentally, Plato's idea that physical reality consists of imperfect imitations of abstractions seems an unnecessarily asymmetrical stance nowadays. Like Plato, we still study abstractions for their own sake. But in post-Galilean science, and in the theory of virtual reality, we also regard abstractions as means of understanding real or artificial *physical* entities, and in that context we take it for granted that the abstractions are nearly always *approximations* to the true physical situation. So, whereas Plato thought of Earthly circles in the sand as approximations to true, mathematical circles, a modern physicist would regard a mathematical circle as a bad approximation to the real shapes of planetary orbits, atoms and other physical things.

Given that there will always be a possibility that the virtual-reality generator or its user interface will go wrong, can a virtual-reality rendering of a Euclidean circle really be said to achieve perfection, up to the standards of mathematical certainty? It can. No one claims that mathematics itself is free from *that* sort of uncertainty. Mathematicians can miscalculate, mis-remember axioms, introduce misprints into their accounts of their own work, and so on. The claim is that, *apart from blunders*, their conclusions are infallible. Similarly, the virtual-reality generator, when it was working properly according to its design specifications, would render a perfect Euclidean circle perfectly.

A similar objection would be that we can never tell for sure how the virtual-reality generator will behave under the control of a given program, because that depends on the functioning of the machine and ultimately on the laws of physics. Since we cannot know the laws of physics for sure, we cannot know for sure that the machine is genuinely rendering Euclidean geometry. But again, no one denies that unforeseen physical phenomena – whether they result from unknown laws of physics or merely from brain disease or trick ink – could mislead a mathematician. But if the laws of physics are in relevant respects as we think they are, then the virtual-reality generator can do its job perfectly, even though we cannot be certain that it is doing so. We must be careful here to distinguish between two issues: whether *we can know* that the virtual-reality machine is rendering a perfect circle; and whether it is *in fact* rendering one. We can never know for sure, but that need not detract one iota from the perfection of the circle that the machine actually renders. I shall return to this crucial distinction – between perfect knowledge (certainty) about an entity, and the entity itself being 'perfect' – in a moment.

Suppose that we deliberately modify the Euclidean geometry program so that the virtual-reality generator will still render circles quite well, but less than perfectly. Would we be unable to infer *anything* about perfect circles by experiencing this imperfect rendering? That would depend entirely on whether we knew in what respects the program had been altered. If we did know, we could work out with certainty (apart from blunders, etc.) which aspects of the experiences we had within the machine would faithfully represent perfect circles, and which would not. And in that case the knowledge we gained there would be just as reliable as any we gained when we were using the correct program.

When we *imagine* circles we are doing just this sort of virtual-reality rendering within our own brains. The reason why this is not a useless way of thinking about perfect circles is that we are able to form accurate theories about what properties our imagined circles do or do not share with perfect ones.

Using a perfect virtual-reality rendering, we might experience six identical circles touching the edge of another identical circle in a plane without overlapping. This experience, under those circumstances, would amount to a rigorous proof that such a pattern is possible, because the geometrical properties of the rendered shapes would be absolutely identical with those of the abstract shapes. But this sort of 'hands-on' interaction with perfect shapes is not capable of yielding *every* sort of knowledge of Euclidean geometry. Most of the interesting theorems refer not to one geometrical pattern but to infinite classes of patterns. For example, the sum of the angles of any Euclidean triangle is 180°. We can measure particular triangles with perfect accuracy in virtual reality, but even in virtual reality we cannot measure all triangles, and so we cannot verify the theorem.

How do we verify it? We prove it. A proof is traditionally defined as a sequence of statements satisfying self-evident rules of inference, but what does the 'proving' *process* amount to physically? To prove a statement about infinitely many triangles at once, we examine certain physical objects – in this case symbols – which have properties in common with whole classes of triangles. For example, when, under appropriate circumstances, we observe the symbols '$\triangle ABC \equiv \triangle DEF$' (i.e. 'triangle ABC is congruent to triangle DEF'), we conclude that a whole class of triangles that we have defined in a particular way always have the same shape as corresponding triangles in another class which we have defined differently. The 'appropriate circumstances' that give this conclusion the status of proof are, in physical terms, that the symbols appear on a page underneath other symbols (some of which represent axioms of Euclidean geometry) and that the pattern in which the symbols appear conforms to certain rules, namely the rules of inference.

But which rules of inference should we use? This is like asking how we should program the virtual-reality generator to make it render the world of Euclidean geometry. The answer is that we must use rules of inference which, to the best of our understanding, will cause our symbols to behave, in the relevant ways, like the abstract entities they denote. How can we be sure that they will? We cannot. Suppose that some critics object to our rules of inference because they think that our symbols will behave differently from the abstract entities. We cannot appeal to the authority of Aristotle or Plato, nor can we prove that our rules of inference are infallible (quite apart from Gödel's theorem, this would lead to an infinite regress, for we should first have to prove that the method of proof that we used was itself valid). Nor can we haughtily tell the critics that there must be something wrong with their intuition, because *our* intuition says that the symbols will mimic the abstract entities perfectly. All we can do is explain. We must explain why we think that, under the circumstances, the symbols will behave in the desired way under our proposed rules. And the critics can explain why they favour a rival theory. A disagreement over two such theories is, in part, a disagreement about the observable behaviour of physical objects. Such disagreements can be addressed by the normal methods of science. Sometimes they can be readily resolved; sometimes not. Another cause of such a disagreement could be a conceptual clash about the nature of the abstract entities themselves. Then again, it is a matter of rival explanations, this time about abstractions rather than physical objects. Either we could come to a common understanding with our critics, or we could agree that we were discussing two different abstract objects, or we could fail to agree. There are no guarantees. Thus, contrary to the traditional belief, it is not the case that disputes within mathematics can always be resolved by purely procedural means.

A conventional symbolic proof seems at first sight to have quite a different character from the 'hands-on' virtual-reality sort of proof. But we see now that they are related in the way that computations are to physical experiments. Any physical experiment can be regarded as a computation, and any computation is a physical experiment. In both sorts of proof, physical entities (whether in virtual reality or not) are manipulated according to rules. In both cases the physical entities represent the abstract entities of interest. And in both cases the reliability of the proof depends on the truth of the theory that physical and abstract entities do indeed share the appropriate properties.

We can also see from the above discussion that proof is a physical *process*. In fact, a proof is a type of computation. 'Proving' a proposition means performing a computation which, if one has done it correctly, establishes that the proposition is true. When we use the word 'proof' to denote an *object*, such as an ink-on-paper text, we mean that the object can be used as a program for recreating a computation of the appropriate kind.

It follows that neither the theorems of mathematics, nor the process of mathematical proof, nor the experience of mathematical intuition, confers any certainty. Nothing does. Our mathematical knowledge may, just like our scientific knowledge, be deep and broad, it may be subtle and wonderfully explanatory, it may be uncontroversially accepted; but it cannot be certain. No one can guarantee that a proof that was previously thought to be valid will not one day turn out to contain a profound misconception, made to seem natural by a previously unquestioned 'self-evident' assumption either about the physical world, or about the abstract world, or about the way in which some physical and abstract entities are related.

It was just such a mistaken, self-evident assumption that caused geometry itself to be mis-classified as a branch of mathematics for over two millennia, from about 300 BC when Euclid wrote his *Elements*, to the nineteenth century (and indeed in most dictionaries and schoolbooks to this day). Euclidean geometry formed part of every mathematician's intuition. Eventually some mathematicians began to doubt that one in particular of Euclid's axioms was self-evident (the so-called 'parallel axiom'). They did not, at first, doubt that this axiom was true. The great German mathematician Carl Friedrich Gauss is said to have been the first to put it to the test. The parallel axiom is required in the proof that the angles of a triangle add up to 180°. Legend has it that, in the greatest secrecy (for fear of ridicule), Gauss placed assistants with lanterns and theodolites at the summits of three hills, the vertices of the largest triangle he could conveniently measure. He detected no deviation from Euclid's predictions, but we now know that that was only because his instruments were not sensitive enough. (The vicinity of the Earth happens to be rather a tame place geometrically.) Einstein's general theory of relativity included a new theory of geometry that contradicted Euclid's and has been vindicated by experiment. The angles of a real triangle really do *not* necessarily add up to 180°: the true total depends on the gravitational field within the triangle.

A very similar mis-classification has been caused by the fundamental mistake that mathematicians since antiquity have been making about the very nature of their subject, namely that mathematical knowledge is more certain than any other form of knowledge. Having made that mistake, one has no choice but to classify proof theory as part of mathematics, for a mathematical theorem could not be certain if the theory that justifies its method of proof were itself uncertain. But as we have just seen, proof theory is not a branch of mathematics – it is a science. Proofs are not abstract. There is no such thing as abstractly proving something, just as there is no such thing as abstractly calculating or computing something. One can of course define a class of abstract entities and call them 'proofs', but those 'proofs' cannot verify mathematical statements because no one can see them. They cannot persuade anyone of the truth of a proposition, any more than an abstract virtual-reality generator that does not physically exist can persuade people that they are in a different environment, or an abstract computer can factorize a number for us. A mathematical 'theory of proofs' would have no bearing on which mathematical truths can or cannot be proved in reality, just as a theory of abstract 'computation' has no bearing on what mathematicians – or anyone else – can or cannot calculate in reality, unless there is a separate, empirical reason for believing that the abstract 'computations' in the theory resemble real computations. Computations, including the special computations that qualify as proofs, are physical processes. Proof theory is about how to ensure that those processes correctly mimic the abstract entities they are intended to mimic.

Gödel's theorems have been hailed as 'the first new theorems of pure logic for two thousand years'. But that is not so: Gödel's theorems are about what can and cannot be proved, and proof is a physical process. Nothing in proof theory is a matter of logic alone. The new way in which Gödel managed to prove general assertions about proofs depends on certain assumptions about which physical processes can or cannot represent an abstract fact in a way that an observer can detect and be convinced by. Gödel distilled such assumptions into his explicit and tacit justification of his results. His results were self-evidently justified, not because they were 'pure logic' but because mathematicians found the assumptions self-evident.

One of Gödel's assumptions was the traditional one that a proof can have only a finite number of steps. The intuitive justification of this assumption is that we are finite beings and could never grasp a literally infinite number of assertions. This intuition, by the way, caused many mathematicians to worry when, in 1976, Kenneth Appel and Wolfgang Haken used a computer to prove the famous 'four-colour conjecture' (that using only four different colours, any map drawn in a plane can be coloured so that no two adjacent regions have the same colour). The program required hundreds of hours of computer time, which meant that the steps of the proof, if written down, could not have been read, let alone recognized as self-evident, by a human being in many lifetimes. 'Should we take the computer's word for it that the four-colour conjecture is proved?', the sceptics wondered – though it had never occurred to them to catalogue all the firings of all the neurons in their own brains when they accepted a relatively 'simple' proof.

The same worry may seem more justified when applied to a putative proof with an infinite number of steps. But what is a 'step', and what is 'infinite'? In the fifth century BC Zeno of Elea concluded, on the basis of a similar intuition, that Achilles will never overtake the tortoise if the tortoise has a head start. After all, by the time Achilles reaches the point where the tortoise is now, it will have moved on a little. By the time he reaches *that* point, it will have moved a little further, and so on *ad infinitum*. Thus the 'catching-up' procedure requires Achilles to perform an infinite number of catching-up steps, which as a finite being he supposedly cannot do. But what Achilles can do cannot be discovered by pure logic. It depends entirely on what the governing laws of physics say he can do. And if those laws say he will overtake the tortoise, then overtake it he will. According to classical physics, catching up requires an infinite number of steps of the form 'move to the tortoise's present location'. In that sense it is a computationally infinite operation. Equivalently, considered as a proof that one abstract quantity becomes larger than another when a given set of operations is applied, it is a proof with an infinite number of steps. But the relevant laws designate it as a physically finite process – and that is all that counts.

Gödel's intuition about steps and finiteness does, as far as we know, capture real physical constraints on the process of proof. Quantum theory requires discrete steps, and none of the known ways in which physical objects can interact would allow for an infinite number of steps to precede a measurable conclusion. (It might, however, be possible for an infinite number of steps to be completed in the whole history of the universe – as I shall explain in Chapter 14.) Classical physics would not have conformed to these intuitions if (impossibly) it had been true. For example, the continuous motion of classical systems would have allowed for 'analogue' computation which did not proceed in steps and which had a substantially different repertoire from the universal Turing machine. Several examples are known of contrived classical laws under which an infinite amount of computation (infinite, that is, by Turing-machine or quantum-computer standards) could be performed by physically finite methods. Of course, classical physics is incompatible with the results of countless experiments, so it is rather artificial to speculate on what the 'actual' classical laws of physics 'would have been'; but what these examples show is that one cannot *prove*, independently of any knowledge of physics, that a proof must consist of finitely many steps. The same considerations apply to the intuition that there must be finitely many rules of inference, and that these must be 'straightforwardly applicable'. None of these requirements is meaningful in the abstract: they are physical requirements. Hilbert, in his influential essay 'On the Infinite', contemptuously ridiculed the idea that the 'finite-number-of-steps' requirement is a substantive one. But the above argument shows that he was mistaken: it is substantive, and it follows only from his and other mathematicians' *physical* intuition.

At least one of Gödel's intuitions about proof turns out to have been mistaken; fortunately, it happens not to affect the proofs of his theorems. He inherited it intact from the prehistory of Greek mathematics, and it remained unquestioned by every generation of mathematicians until it was proved false in the 1980s by discoveries in the quantum theory of computation. It is the intuition that a proof is a particular type of *object*, namely a sequence of statements that obey rules of inference. I have already argued that a proof is better regarded not as an object but as a process, a type of computation. But in the classical theory of proof or computation this makes no fundamental difference, for the following reason. If we can go through the process of a proof, we can, with only a moderate amount of extra effort, keep a record of everything relevant that happens during that process. That record, a physical object, will constitute a proof in the sequence-of-statements sense. And conversely, if we have such a record we can read through it, checking that it satisfies the rules of inference, and in the process of doing so we shall have proved the conclusion. In other words, in the classical case, converting between proof processes and proof objects is always a tractable task.

Now consider some mathematical calculation that is intractable on all classical computers, but suppose that a quantum computer can easily perform it using interference between, say, 10^{500} universes. To make the point more clearly, let the calculation be such that the answer (unlike the result of a factorization) cannot be tractably verified once we have it. The process of programming a quantum computer to perform such a computation, running the program and obtaining a result, constitutes a proof that the mathematical calculation has that particular result. But now there is no way of keeping a record of everything that happened during the proof process, because most of it happened in other universes, and measuring the computational state would alter the interference properties and so invalidate the proof. So creating an old-fashioned proof *object* would be infeasible; moreover, there is not remotely enough material in the universe as we know it to make such an object, since there would be vastly more steps in the proof than there are atoms in the known universe. This example shows that because of the possibility of quantum computation, the two notions of proof are not equivalent. The intuition of a proof as an object does not capture all the ways in which a mathematical statement may in reality be proved.

Once again, we see the inadequacy of the traditional mathematical method of deriving certainty by trying to strip away every possible source of ambiguity or error from our intuitions until only self-evident truth remains. That is what Gödel had done. That is what Church, Post and especially Turing had done when trying to intuit their universal models for computation. Turing hoped that his abstracted-paper-tape model was so simple, so transparent and well defined, that it would not depend on any assumptions about physics that could conceivably be falsified, and therefore that it could become the basis of an abstract theory of computation that was independent of the underlying physics. 'He thought,' as Feynman once put it, 'that he understood paper.' But he was mistaken. Real, quantum-mechanical paper is wildly different from the abstract stuff that the Turing machine uses. The Turing machine is entirely classical, and does not allow for the possibility that the paper might have different symbols written on it in different universes, and that those might interfere with one another. Of course, it is impractical to detect interference between different states of a paper tape. But the point is that Turing's intuition, because it included false assumptions from classical physics, caused him to abstract away some of the *computational* properties of his hypothetical machine, the very properties he intended to keep. That is why the resulting model of computation was incomplete.

That mathematicians throughout the ages should have made various mistakes about matters of proof and certainty is only natural. The present discussion should lead us to expect that the current view will not last for ever, either. But the confidence with which mathematicians have blundered into these mistakes and their inability to acknowledge even the possibility of error in these matters are, I think, connected with an ancient and widespread confusion between the *methods* of mathematics and its *subject-matter*. Let me explain. Unlike the relationships between physical entities, relationships between abstract entities are independent of any contingent facts and of any laws of physics. They are determined absolutely and objectively by the autonomous properties of the abstract entities themselves. Mathematics, the study of these relationships and properties, is therefore the study of *absolutely necessary truths*. In other words, the truths that mathematics *studies* are absolutely certain. But that does not mean that our knowledge of those necessary truths is itself certain, nor does it mean that the methods of mathematics confer necessary truth on their conclusions. After all, mathematics also studies falsehoods and paradoxes. And that does not mean that the conclusions of such a study are necessarily false or paradoxical.

Necessary truth is merely the *subject-matter* of mathematics, not the reward we get for doing mathematics. The objective of mathematics is not, and cannot be, mathematical certainty. It is not even mathematical truth, certain or otherwise. It is, and must be, mathematical explanation.

Why, then, does mathematics work as well as it does? Why does it lead to conclusions which, though not certain, can be accepted and applied unproblematically for millennia at least? Ultimately the reason is that *some* of our knowledge of the physical world is also that reliable and uncontroversial. And when we understand the physical world sufficiently well, we also understand which physical objects have properties in common with which abstract ones. But in principle the reliability of our knowledge of mathematics remains subsidiary to our knowledge of physical reality. Every mathematical proof depends absolutely for its validity on our being right about the rules that govern the behaviour of some physical objects, be they virtual-reality generators, ink and paper, or our own brains.

So mathematical intuition is a species of physical intuition. Physical intuition is a set of rules of thumb, some perhaps inborn, many built up in childhood, about how the physical world behaves. For example, we have intuitions that there are such things as physical objects, and that they have definite attributes such as shape, colour, weight and position in space, some of which exist even when the objects are unobserved. Another is that there is a physical variable – time – with respect to which attributes change, but that nevertheless objects can retain their identity over time. Another is that objects interact, and that this can change some of their attributes. Mathematical intuition concerns the way in which the physical world can display the properties of abstract entities. One such intuition is that of an abstract law, or at least an explanation, that underlies the behaviour of objects. The intuition that space admits closed surfaces that separate an 'inside' from an 'outside' may be refined into the mathematical intuition of a *set*, which partitions everything into members and non-members of the set. But further refinement by mathematicians (starting with Russell's refutation of Frege's set theory) has shown that this intuition ceases to be accurate when the sets in question contain 'too many' members (too large a degree of infinity of members).

Even if any physical or mathematical intuition were inborn, that would not confer any special authority upon it. Inborn intuition cannot be taken as a surrogate for Plato's 'memories' of the world of Forms. For it is a commonplace observation that many of the intuitions built into human beings by accidents of evolution are simply false. For example, the human eye and its controlling software implicitly embody the false theory that yellow light consists of a mixture of red and green light (in the sense that yellow light gives us exactly the same sensation as a mixture of red light and green light does). In reality, all three types of light have different frequencies and cannot be created by mixing light of other frequencies. The fact that a mixture of red and green light appears to us to be yellow light has nothing whatever to do with the properties of light, but is a property of our eyes. It is the result of a design compromise that occurred at some time during our distant ancestors' evolution. It is just possible (though I do not believe it) that Euclidean geometry or Aristotelian logic are somehow built into the structure of our brains, as the philosopher Immanuel Kant believed. But that would not logically imply that they were true. Even in the still more implausible event that we have inborn intuitions that we are constitutionally unable to shake off, such intuitions would still not be necessary truths.

The fabric of reality, then, does have a more unified structure than would have been possible if mathematical knowledge had been verifiable with certainty, and hence hierarchical, as has traditionally been assumed. Mathematical entities are part of the fabric of reality because they are complex and autonomous. The sort of reality they form is in some ways like the realm of abstractions envisaged by Plato or Penrose: although they are by definition intangible, they exist objectively and have properties that are independent of the laws of physics. However, it is physics that allows us to gain knowledge of this realm. And it imposes stringent constraints. Whereas everything in physical reality is comprehensible, the comprehensible mathematical truths are precisely the infinitesimal minority which happen to correspond exactly to some physical truth – like the fact that if certain symbols made of ink on paper are manipulated in certain ways, certain other symbols appear. That is, they are the truths that can be rendered in virtual reality. We have no choice but to assume that the incomprehensible mathematical entities are real too, because they appear inextricably in our explanations of the comprehensible ones.

There are physical objects – such as fingers, computers and brains – whose behaviour can model that of certain abstract objects. In this way the fabric of physical reality provides us with a window on the world of abstractions. It is a very narrow window and gives us only a limited range of perspectives. Some of the structures that we see out there, such as the natural numbers or the rules of inference of classical logic, seem to be important or 'fundamental' to the abstract world, in the same way as deep laws of nature are fundamental to the physical world. But that could be a misleading appearance. For what we are really seeing is only that some abstract structures are fundamental *to our understanding* of abstractions. We have no reason to suppose that those structures are objectively significant in the abstract world. It is merely that some abstract entities are nearer and more easily visible from our window than others.

TERMINOLOGY

mathematics The study of absolutely necessary truths.

proof A way of establishing the truth of mathematical propositions.

(Traditional definition:) A sequence of statements, starting with some premises and ending with the desired conclusion, and satisfying certain 'rules of inference'.

(Better definition:) A computation that models the properties of some abstract entity, and whose outcome establishes that the abstract entity has a given property.

mathematical intuition (Traditionally:) An ultimate, self-evident source of justification for mathematical reasoning.

(Actually:) A set of theories (conscious and unconscious) about the behaviour of certain physical objects whose behaviour models that of interesting abstract entities.

intuitionism The doctrine that all reasoning about abstract entities is untrustworthy except where it is based on direct, self-evident intuition. This is the mathematical version of solipsism.

Hilbert's tenth problem To 'establish once and for all the certitude of mathematical methods' by finding a set of rules of inference sufficient for all valid proofs, and then proving those rules consistent by their own standards.

Gödel's incompleteness theorem A proof that Hilbert's tenth problem cannot be solved. For any set of rules of inference, there are valid proofs not designated as valid by those rules.

SUMMARY

Abstract entities that are complex and autonomous exist objectively and are part of the fabric of reality. There exist logically necessary truths about these entities, and these comprise the subject-matter of mathematics. However, such truths cannot be known with certainty. Proofs do not confer certainty upon their conclusions. The validity of a particular form of proof depends on the truth of our theories of the behaviour of the objects with which we perform the proof. Therefore mathematical knowledge is inherently derivative, depending entirely on our knowledge of physics. The comprehensible mathematical truths are precisely the infinitesimal minority which can be rendered in virtual reality. But the incomprehensible mathematical entities (e.g. Cantgotu environments) exist too, because they appear inextricably in our explanations of the comprehensible ones.

I have said that computation always was a quantum concept, because classical physics was incompatible with the intuitions that formed the basis of the classical theory of computation. The same thing is true of time. Millennia before the discovery of quantum theory, time was the first quantum concept.

11

Time: The First Quantum Concept

Like as the waves make towards the pebbled shore,
So do our minutes hasten to their end;
Each changing place with that which goes before,
In sequent toil all forwards do contend.
William Shakespeare (Sonnet 60)

Even though it is one of the most familiar attributes of the physical world, time has a reputation for being deeply mysterious. Mystery is part of the very concept of time that we grow up with. St Augustine, for example, said:

What then is time? If no one asks me, I know; if I wish to explain it to one who asks, I know not. (*Confessions*)

Few people think that distance is mysterious, but everyone knows that time is. And all the mysteries of time stem from its basic, common-sense attribute, namely that the present moment, which we call 'now', is not fixed but moves continuously in the future direction. This motion is called the *flow* of time.

We shall see that there is no such thing as the flow of time. Yet the idea of it is pure common sense. We take it so much for granted that it is assumed in the very structure of our language. In *A Comprehensive Grammar of the English Language*, Randolph Quirk and his co-authors explain the common-sense concept of time with the aid of the diagram shown in Figure 11.1. Each point on the line represents a particular, fixed moment. The triangle '▽' indicates where the 'continuously moving point, the present moment', is located on the line. It is supposed to be moving from left to right. Some people, like Shakespeare in the sonnet quoted above, think of particular events as being 'fixed', and the line itself as moving past them (from right to left in Figure 11.1), so that moments from the future sweep past the present moment to become past moments.

What do we mean by 'time can be thought of as a line?' We mean that just as a line can be thought of as a sequence of points at different positions, so any moving or changing object can be thought of as a sequence of motionless 'snapshot' versions of itself, one at each moment. To say that each point on the line represents a particular moment is to say that we can imagine all the snapshots stacked together along the line, as in Figure 11.2. Some of them show the rotating arrow as it was in the past, some show it as it will be in the future, and one of them – the one to which the moving ∇ is currently pointing – shows the arrow as it is now, though a moment later that particular version of the arrow will be in the past because the ∇ will have moved on. The instantaneous versions of an object collectively *are* the moving object in much the way that a sequence of still pictures projected onto a screen collectively *are* a moving picture. None of them, individually, ever changes. Change consists of their being designated ('illuminated') in sequence by the moving ∇ (the 'movie projector') so that, one by one, they take it in turn to be in the present.

Grammarians nowadays try not to make value-judgements about how language is used; they try only to record, analyse and

...'time can be thought of as a line (theoretically, of infinite length) on which is located, as a continuously moving point, the present moment. Anything ahead of the present moment is in the future, and anything behind it is in the past.'

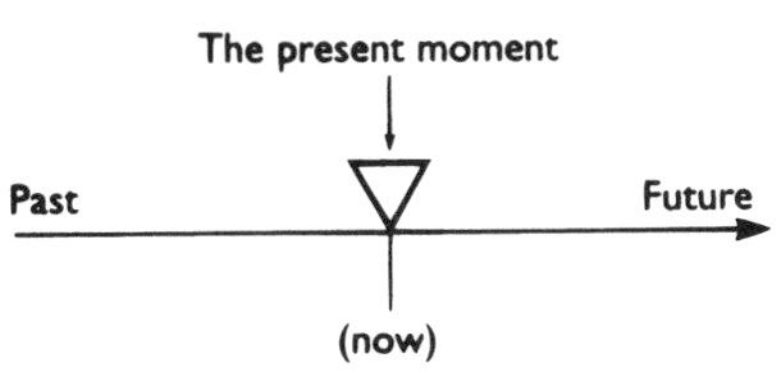

FIGURE 11.1 *The common-sense concept of time that is assumed in the English language (based on Quirk* et al., A Comprehensive Grammar of the English Language, *p. 175).*

understand it. Therefore Quirk *et al.* are in no way to blame for the quality of the theory of time that they describe. They do not claim that it is a good theory. They claim only, and I think quite correctly, that it is *our* theory. Unfortunately it is not a good theory. To put it bluntly, the reason why the common-sense theory of time is inherently mysterious is that it is inherently nonsensical. It is not just that it is factually inaccurate. We shall see that, even in its own terms, it does not make sense.

This is perhaps surprising. We have become used to modifying our common sense to conform to scientific discoveries. Common sense frequently turns out to be false, even badly false. But it is unusual for common sense to be *nonsense* in a matter of everyday experience. Yet that is what has happened here.

Consider Figure 11.2 again. It illustrates the motion of two entities. One of them is a rotating arrow, shown as a sequence of snapshots. The other is the moving 'present moment', sweeping through the picture from left to right. But the motion of the present moment is not shown in the picture as a sequence of snapshots. Instead, one particular moment is singled out by the ▽, highlighted in darker lines and uniquely labelled '(now)'. Thus, even though 'now' is said by the caption to be moving across the picture, only one snapshot of it, at one particular moment, is shown.

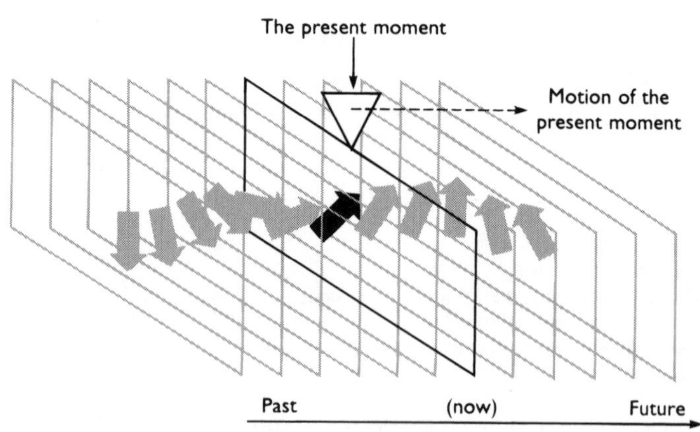

FIGURE 11.2 *A moving object as a sequence of 'snapshots', which become the present moment one by one.*

Why? After all, the whole point of this picture is to show what happens over an extended period, not just at one moment. If we had wanted the picture to show only one moment, we need not have bothered to show more than one snapshot of the rotating arrow either. The picture is supposed to illustrate the common-sense theory that any moving or changing object is a sequence of snapshots, one for each moment. So if the ∇ is moving, why do we not show a sequence of snapshots of it too? The single snapshot shown must be only one of many that would exist if this were a true description of how time works. In fact, the picture is positively misleading as it stands: it shows the ∇ *not* moving, but rather coming into existence at a particular moment and then immediately ceasing to exist. If that were so, it would make 'now' a *fixed* moment. It makes no difference that I have added a label 'Motion of the present moment', and a dashed arrow to indicate that the ∇ is moving to the right. What the picture itself shows, and what Quirk *et al.*'s diagram (Figure 11.1) also shows, is the ∇ never reaching any moment other than the highlighted one.

At best, one could say that Figure 11.2 is a hybrid picture which perversely illustrates motion in two different ways. In regard to the moving arrow it illustrates the common-sense theory of time. But it merely *states* that the present moment is moving, while illustrating it as *not* moving. How should we alter the picture so that it will illustrate the common-sense theory of time in regard to the motion of the present moment as well as the motion of the arrow? By including more snapshots of the '∇', one for each moment, each indicating where 'now' is at that moment. And where is that? Obviously, at each moment, 'now' is that moment. For example, at midnight the '∇' must point to the snapshot of the arrow taken at midnight; at 1.00 a.m. it must point to the 1.00 a.m. snapshot, and so on. Therefore the picture should look like Figure 11.3.

This amended picture illustrates *motion* satisfactorily, but we are now left with a severely pared-down concept of time. The common-sense idea that a moving object is a sequence of instantaneous versions of itself remains, but the other common-sense idea – of the flow of time – has gone. In this picture there is no 'continuously moving point, the present moment', sweeping through the fixed moments one by one. There is no process by which any fixed moment starts out in the future, becomes the present and is then relegated to the past. The multiple instances of the symbols ▽ and '(now)' no longer distinguish one moment from others, and are therefore superfluous. The picture would illustrate the motion of the rotating arrow just as well if they were removed.

So there is no single 'present moment', except subjectively. From the point of view of an observer at a particular moment, that moment is indeed singled out, and may uniquely be called 'now' by that observer, just as any position in space is singled out as 'here' from the point of view of an observer at that position. But objectively, no moment is privileged as being more 'now' than the others, just as no position is privileged as being more 'here' than other positions. The subjective 'here' may move through space, as the observer moves. Does the subjective 'now' likewise move through time? Are Figures 11.1 and 11.2 correct after all, in that they illustrate time from the point of view of an observer at a particular moment? Certainly not. Even subjectively, 'now' *does not move through time*. It is often said that the present *seems* to be moving forwards in time because the present is defined only relative to our consciousness, and our consciousness is sweeping forwards through the moments. But our consciousness does not,

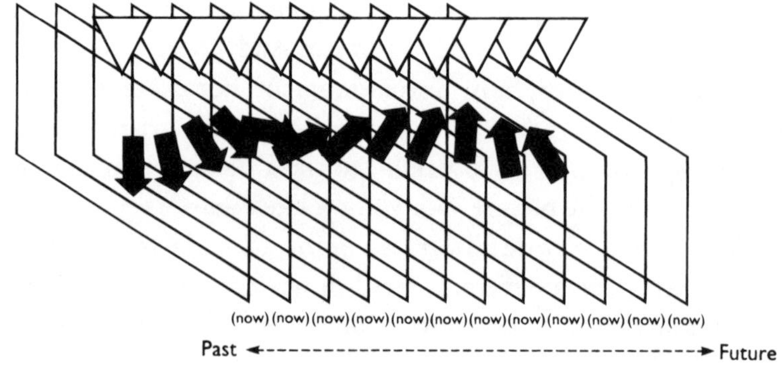

FIGURE 11.3 *At each moment, 'now' is that moment.*

and could not, do that. When we say that our consciousness 'seems' to pass from one moment to the next we are merely paraphrasing the common-sense theory of the flow of time. But it makes no more sense to think of a single 'moment of which we are conscious' moving from one moment to another than it does to think of a single present moment, or anything else, doing so. *Nothing* can move from one moment to another. To exist at all at a particular moment means to exist there for ever. Our consciousness exists at *all* our (waking) moments.

Admittedly, different snapshots of the observer perceive different moments as 'now'. But that does not mean that the observer's consciousness – or any other moving or changing entity – moves through time as the present moment is supposed to. The various snapshots of the observer do not take it in turns to be in the present. They do not take it in turns to be conscious of their present. They are all conscious, and subjectively they are all in the present. Objectively, there is no present.

We do not experience time flowing, or passing. What we experience are differences between our present perceptions and our present memories of past perceptions. We interpret those differences, correctly, as evidence that the universe changes with time. We also interpret them, incorrectly, as evidence that our consciousness, or the present, or something, moves through time.

If the moving present capriciously stopped moving for a day or two, and then started to move again at ten times its previous speed, what would we be conscious of? Nothing special – or rather, that question makes no sense. There is nothing there that could move, stop or flow, nor could anything be meaningfully called the 'speed' of time. Everything that exists in time is supposed to take the form of unchanging snapshots arrayed along the time-line. That includes the conscious experiences of all observers, including their mistaken intuition that time is 'flowing'. They may imagine a 'moving present' travelling along the line, stopping and starting, or even going backwards or ceasing to exist altogether. But imagining it does not make it happen. Nothing can move along the line. Time cannot flow.

The idea of the flow of time really presupposes the existence of a second sort of time, outside the common-sense sequence-of-moments time. If 'now' really moved from one of the moments to another, it would have to be with respect to this *exterior* time. But taking that seriously leads to an infinite regress, for we should then have to imagine the exterior time itself as a succession of moments, with its own 'present moment' that was moving with respect to a still more exterior time – and so on. At each stage, the flow of time would not make sense unless we attributed it to the flow of an exterior time, *ad infinitum*. At each stage, we would have a concept that made no sense; and the whole infinite hierarchy would make no sense either.

The origin of this sort of mistake is that we are accustomed to time being a framework exterior to any physical entity we may be considering. We are used to imagining any physical object as potentially changing, and so existing as a sequence of versions of itself at different moments. But the sequence of moments itself, in pictures like Figures 11.1–11.3, is an exceptional entity. It does not exist within the framework of time – it *is* the framework of time. Since there is no time outside it, it is incoherent to imagine it changing or existing in more than one consecutive version. This makes such pictures hard to grasp. The picture itself, like any other physical object, does exist over a period of time and does consist of multiple versions of itself. But what the picture *depicts* – namely, the sequence of versions of something – exists in only one version. No accurate picture of the framework of time can be a moving or changing picture. It must be static. But there is an inherent psychological difficulty in taking this on board. Although the picture is static, we cannot understand it statically. It shows a sequence of moments simultaneously on the page, and in order to relate that to our experience the focus of our attention must move along the sequence. For example, we might look at one snapshot, and take it to represent 'now', and a moment later look at a snapshot to the right of it and think of that as representing the new 'now'. Then we tend to confuse the genuine motion of our focus of attention across the mere *picture*, with the impossible motion of something through real moments. It is easily done.

But there is more to this problem than the difficulty of *illustrating* the common-sense theory of time. The theory itself contains a substantive and deep equivocation: it cannot make up its mind whether the present is, objectively, a single moment or many – and hence, for example, whether Figure 11.1 depicts one moment or many. Common sense wants the present to be a single moment so as to allow the flow of time – to allow the present to sweep through the moments from past to future. But common sense also wants time to be a sequence of moments, with all motion and change consisting of differences between versions of an entity at different moments. And that means that the moments are themselves unchanging. So a particular moment cannot become the present, or cease to be the present, for these would be changes. Therefore the present cannot, objectively, be a single moment.

The reason why we cling to these two incompatible concepts – the moving present and the sequence of unchanging moments – is that we need them both, or rather, that we think we do. We continually invoke both of them in everyday life, albeit never quite in the same breath. When we are *describing* events, saying when things happen, we think in terms of a sequence of unchanging moments; when we are *explaining* events as causes and effects of each other, we think in terms of the moving present.

For example, in saying that Faraday discovered electromagnetic induction 'in 1831' we are assigning that event to a certain range of moments. That is, we are specifying on which set of snapshots, in the long sheaf of snapshots of world history, that discovery is to be found. No flow of time is involved when we say *when* something happened, any more than a 'flow of distance' is involved if we say *where* it happened. But as soon as we say *why* something happened, we invoke the flow of time. If we say that we owe our electric motors and dynamos in part to Faraday, and that the repercussions of his discovery are being felt to this day, we have in mind a picture of the repercussions beginning in 1831 and sweeping consecutively through all the moments of the rest of the nineteenth century, and then reaching the twentieth century and causing things like power stations to come into existence there. If we are not careful, we think of the twentieth century as initially 'not yet affected' by the momentous event of 1831, and then being 'changed' by the repercussions as they sweep past on their way to the twenty-first century and beyond. But usually we are careful, and we avoid that incoherent thought by never using the two parts of the common-sense theory of time simultaneously. Only when we think about time itself do we do that, and then we marvel at the mystery of it all! Perhaps 'paradox' is a better word than mystery, for we have here a blatant conflict between two apparently self-evident ideas. They cannot both be true. We shall see that neither is true.

Our theories of physics are, unlike common sense, coherent, and they first achieved this by dropping the idea of the flow of time. Admittedly, physicists *talk* about the flow of time just as everyone else does. For example, in his book *Principia*, in which he set out the principles of Newtonian mechanics and gravitation, Newton wrote:

> Absolute, true, and mathematical time, of itself, and from its own nature, flows equably without relation to anything external.

But Newton wisely makes no attempt to translate his assertion that time flows into mathematical form, or to derive any conclusion from it. None of Newton's physical theories refers to the flow of time, nor has any subsequent physical theory referred to, or been compatible with, the flow of time.

So why did Newton think it necessary to say that time 'flows equably'? There is nothing wrong with 'equably': one can interpret that as meaning that measurements of time are the same for observers at different positions and in different states of motion. That is a substantive assertion (which, since Einstein, we know to be inaccurate). But it could easily have been stated as I have just stated it, without saying that time flows. I think that Newton was deliberately using the familiar language of time without intending its literal meaning, just as he might have spoken informally of the Sun 'rising'. He needed to convey to the reader embarking on this revolutionary work that there was nothing new or sophisticated in the Newtonian concept of time. The *Principia* assigns to many words, such as 'force' and 'mass', precise technical meanings which are somewhat different from their common-sense ones. But the numbers referred to as 'times' are simply the times of common sense, which we find on clocks and calendars, and the concept of time in the *Principia* is the common-sense one.

Only, it does not flow. In Newtonian physics, time and motion appear much as in Figure 11.3. One minor difference is that I have been drawing successive moments separated from one another, but in all pre-quantum physics that is an approximation because time is a continuum. We must imagine infinitely many, infinitely thin snapshots interpolating continuously between the ones I have drawn. If each snapshot represents everything throughout the whole of space that physically exists at a particular moment, then we can think of the snapshots as being glued together at their faces to form a single, unchangeable block containing everything that happens in space and time (Figure 11.4) – that is, the whole of physical reality. An inevitable shortcoming of this sort of diagram is that the snapshots of space at each moment are shown as being two-dimensional, whereas in reality they are three-dimensional. Each one of them is space at a particular moment. Thus we are treating time as a fourth dimension, analogous to the three dimensions of space in classical

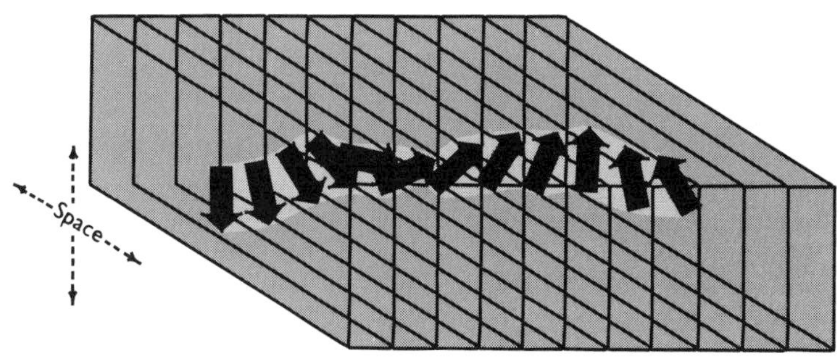

FIGURE 11.4 *Spacetime, considered as successive moments.*

geometry. Space and time, considered together like this as a four-dimensional entity, are called *spacetime*.

In Newtonian physics this four-dimensional geometrical interpretation of time was optional, but under Einstein's theory of relativity it became an indispensable part of the theory. That is because, according to relativity, observers moving at different velocities do not agree about which events are simultaneous. That is, they do not agree about which events should appear on the same snapshot. So they each perceive spacetime as being sliced up in a different way into 'moments'. Nevertheless, if they each stacked their snapshots in the manner of Figure 11.4, the spacetimes they constructed would be identical. Therefore, according to relativity, the 'moments' shown in Figure 11.4 are not objective features of spacetime: they are only one observer's way of perceiving simultaneity. Another observer would draw the 'now' slices at a different angle. So the objective reality behind Figure 11.4, namely the spacetime and its physical contents, could be shown as in Figure 11.5.

Spacetime is sometimes referred to as the 'block universe', because within it the whole of physical reality – past, present and future – is laid out once and for all, frozen in a single four-dimensional block. Relative to spacetime, nothing ever moves. What we call 'moments' are certain slices through spacetime, and when the contents of such slices are different from one another, we call it change or motion through space.

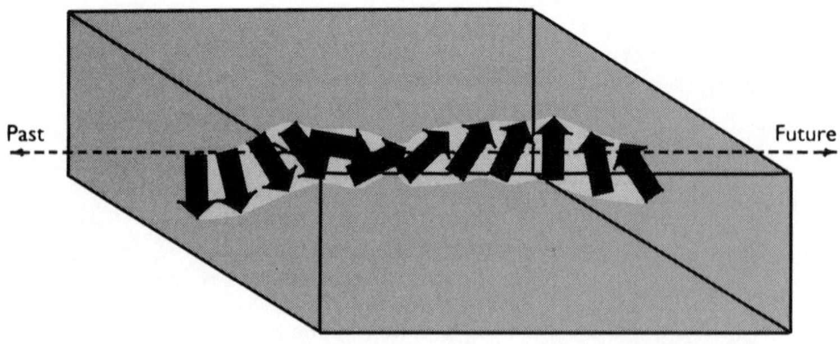

FIGURE 11.5 *Spacetime view of a moving object.*

As I have said, we think of the flow of time in connection with causes and effects. We think of causes as preceding their effects; we imagine the moving present arriving at causes before it arrives at their effects, and we imagine the effects flowing forwards with the present moment. Philosophically, the most important cause-and-effect processes are our conscious decisions and the consequent actions. The common-sense view is that we have *free will*: that we are sometimes in a position to affect future events (such as the motion of our own bodies) in any one of several possible ways, and to choose which shall occur; whereas, in contrast, we are never in a position to affect the past at all. (I shall discuss free will in Chapter 13.) The past is fixed; the future is open. To many philosophers, the flow of time is the process in which the open future becomes, moment by moment, the fixed past. Others say that the alternative events at each moment in the future are *possibilities*, and the flow of time is the process by which, moment by moment, one of these possibilities becomes *actual* (so that, according to those people, the future does not exist at all until the flow of time hits it and turns it into the past). But if the future really is open (and it is!), then that can have nothing to do with the flow of time, for there is no flow of time. In spacetime physics (which is, effectively, all pre-quantum physics, starting with Newton) the future is not open. It is *there*, with definite, fixed contents, just like the past and present. If a particular moment in spacetime were 'open' (in any sense) it would necessarily remain open when it became the present and the past, for moments cannot change.

Subjectively, the future of a given observer may be said to be 'open from that observer's point of view' because one cannot measure or observe one's own future. But openness in that subjective sense does not allow choices. If you have a ticket for last week's lottery, but have not yet found out whether you have won, the outcome is still open from your point of view, even though objectively it is fixed. But, subjectively or objectively, you cannot change it. No causes which have not already affected it can do so any longer. The common-sense theory of free will says that last week, while you still had a choice whether to buy a ticket or not, the future was still objectively open, and you really could have chosen any of two or more options. But that is incompatible with spacetime. So according to spacetime physics, the openness of the future is an illusion, and therefore causation and free will can be no more than illusions as well. We need, and cling to, the belief that the future can be affected by present events, and especially by our choices; but perhaps that is just our way of coping with the fact that we do not know the future. In reality, we make no choices. Even as we think we are considering a choice, its outcome is already there, on the appropriate slice of spacetime, unchangeable like everything else in spacetime, and impervious to our deliberations. It seems that those deliberations themselves are unchangeable and already in existence at their allotted moments before we ever know of them.

To be an 'effect' of some cause means to be affected by that cause – to be changed by it. Thus when spacetime physics denies the reality of the flow of time, it logically cannot accommodate the common-sense notions of cause and effect either. For in the block universe nothing is changeable: one part of spacetime can no more change another than one part of a fixed *three*-dimensional object can change another.

It so happens that all fundamental theories in the era of spacetime physics had the property that given everything that happens before a given moment, the laws of physics determine what happens at all subsequent moments. The property of snapshots being determined by other snapshots is called *determinism*. In Newtonian physics, for instance, if at any moment one knows the positions and velocities of all the masses in an isolated system, such as the solar system, one can in principle calculate (*predict*) where those masses will be at all times thereafter. One can also in principle calculate (*retrodict*) where those masses were at all previous times.

The laws of physics that determine one snapshot from another are the 'glue' that holds the snapshots together as a spacetime. Let us imagine ourselves, magically and impossibly, outside spacetime (and therefore in an external time of our own, independent of that within spacetime). Let us slice spacetime into snapshots of space at each moment as perceived by a particular observer within spacetime, then shuffle the snapshots and glue them together again in a new order. Could we tell, from the outside, that this is not the real spacetime? Almost certainly. For one thing, in the shuffled spacetime physical processes would not be continuous. Objects would instantaneously cease to exist at one point and reappear at another. Second, and more important, the laws of physics would no longer hold. At least, the real laws of physics would no longer hold. There would exist a different set of laws that took the shuffling into account, explicitly or implicitly, and correctly described the shuffled spacetime.

So to us, the difference between the shuffled spacetime and the real one would be gross. But what about the inhabitants? Could they tell the difference? We are getting dangerously close to nonsense here – the familiar nonsense of the common-sense theory of time. But bear with me and we shall skirt around the nonsense. *Of course* the inhabitants could not tell the difference. If they could, they would. They would, for instance, comment on the existence of discontinuities in their world, and publish scientific papers about them – that is, if they could survive in the shuffled spacetime at all. But from our magical vantage-point we can see that they do survive, and so do their scientific papers. We can read those papers, and see that they still contain only observations of the original spacetime. All records within the spacetime of physical events, including those in the memories and perceptions of conscious observers, are identical to those in the original spacetime. We have only shuffled the snapshots, not changed them internally, so the inhabitants still perceive them in the original order.

Thus in terms of real physics – physics as perceived by the spacetime's inhabitants – all this slicing up and re-gluing of spacetime is meaningless. Not only the shuffled spacetime, but even the collection of unglued-together snapshots, is physically identical to the original spacetime. We picture all the snapshots glued together in the right order because this represents the relationships between them that are determined by the laws of physics. A picture of them glued together in a different order would represent the same physical events – the same history – but would somewhat misrepresent the relationships between those events. So the snapshots have an *intrinsic* order, defined by their contents and by the real laws of physics. Any one of the snapshots, together with the laws of physics, not only determines what all the others are, it determines their order, and it determines its own place in the sequence. In other words, each snapshot has a 'time stamp' encoded in its physical contents.

That is how it must be if the concept of time is to be freed of the error of invoking an overarching framework of time that is external to physical reality. The time stamp of a snapshot is the reading on some natural clock that exists within that universe. In some snapshots – the ones containing human civilization, for example – there are actual clocks. In others there are physical variables – such as the chemical composition of the Sun, or of all the matter in space – which can be considered as clocks because they take definite, distinct values on different snapshots, at least over a certain region of spacetime. We can standardize and calibrate them to agree with one another where they overlap.

We can reconstitute the spacetime by using the intrinsic order determined by the laws of physics. We start with any of the snapshots. Then we calculate what the immediately preceding and following snapshots should look like, and we locate those snapshots from the remaining collection and glue them on either side of the original snapshot. Repeating the process builds up the whole spacetime. These calculations are too complex to perform in real life, but they are legitimate in a thought experiment in which we imagine ourselves to be detached from the real, physical world. (Also, strictly speaking, in pre-quantum physics there would be a continuous infinity of snapshots, so the process just described would have to be replaced by a limiting process in which the spacetime is assembled in an infinite number of steps; but the principle is the same.)

The predictability of one event from another does not imply that those events are cause and effect. For example, the theory of electrodynamics says that all electrons carry the same charge. Therefore, using that theory we can – and frequently do – predict the outcome of a measurement on one electron from the outcome of a measurement on another. But neither outcome was *caused* by the other. In fact, as far as we know, the value of the charge on an electron was not caused by any physical process. Perhaps it is 'caused' by the laws of physics themselves (though the laws of physics as we currently know them do not predict the charge on the electron; they merely say that all electrons have the same charge). But in any case, here is an example of events (outcomes of measurements on electrons) that are predictable from one another, but make no causal contribution to one another.

Here is another example. If we observe where one piece of a fully assembled jigsaw puzzle is, and we know the shapes of all the pieces, and that they are interlocked in the proper way, we can predict where all the other pieces are. But that does not mean that the other pieces were *caused* to be where they are by the piece we observed being where it is. Whether such causation is involved depends on how the jigsaw puzzle as a whole got there. If the piece we observed was laid down first, then it was indeed among the causes of the other pieces being where they are. If any other piece was laid down first, then the position of the piece we observed was an *effect* of that, not a cause. But if the puzzle was created by a single stroke of a jigsaw-puzzle-shaped cutter, and has never been disassembled, then none of the positions of the pieces are causes or effects of each other. They were not assembled in any order, but were created simultaneously, in positions such that the rules of the puzzle were already obeyed, which made those positions mutually predictable. Nevertheless, none of them caused the others.

The determinism of physical laws about events in spacetime is like the predictability of a correctly interlocking jigsaw puzzle. The laws of physics determine what happens at one moment from what happens at another, just as the rules of the jigsaw puzzle determine the positions of some pieces from those of others. But, just as with the jigsaw puzzle, whether the events at different moments *cause* one another or not depends on how the moments got there. We cannot tell by looking at a jigsaw puzzle whether it got there by being laid down one piece at a time. But with spacetime we know that it does not make sense for one moment to be 'laid down' after another, for that would be the flow of time. Therefore we know that even though some events can be predicted from others no event in spacetime caused another. Let me stress again that this is all according to pre-quantum physics, in which everything that happens, happens in spacetime. What we are seeing is that spacetime is incompatible with the existence of cause and effect. It is not that people are mistaken when they say that certain physical events are causes and effects of one another, it is just that that intuition is incompatible with the laws of spacetime physics. But that is all right, because spacetime physics is false.

I said in Chapter 8 that two conditions must hold for an entity to be a cause of its own replication: first, that the entity is in fact replicated; and second, that most variants of it, in the same situation, would not be replicated. This definition embodies the idea that a cause is something that makes a difference to its effects, and it also works for causation in general. For X to be a cause of Y, two conditions must hold: first, that X and Y both happen; and second, that Y would not have happened if X had been otherwise. For example, sunlight was a cause of life on Earth because both sunlight and life actually occurred on Earth, and because life would not have evolved in the absence of sunlight.

Thus, reasoning about causes and effects is inevitably also about variants of the causes and effects. One is always saying what *would* have happened if, other things being equal, such and such an event had been different. A historian might make the judgement that '*if* Faraday had died in 1830, *then* technology would have been delayed for twenty years'. The meaning of this judgement seems perfectly clear and, since in fact Faraday did not die in 1830 but discovered electromagnetic induction in 1831, it seems quite plausible too. It is equivalent to saying that the technological progress which did happen was in part caused by Faraday's discovery, and therefore also by his survival. But what does it mean, in the context of spacetime physics, to reason about the future of non-existent events? If there is no such event in spacetime as Faraday's death in 1830, then there is also no such thing as its aftermath. Certainly we can *imagine* a spacetime that contains such an event; but then, since we are only imagining it, we can also imagine that it contains any aftermath we like. We can imagine, for example, that Faraday's death was followed by an *acceleration* of technological progress. We might try to get around this ambiguity by imagining only spacetimes in which, though the event in question is different from that in actual spacetime, the laws of physics are the same. It is not clear what justifies restricting our imagination in this way, but in any case, if the laws of physics are the same then the event in question *could not* have been different, because the laws determine it unambiguously from the previous history. So the previous history would have to be imagined as being different as well. How different? The effect of our imagined variation in history depends critically on what we take 'other things being equal' to mean. And that is irreducibly ambiguous, for there are infinitely many ways of imagining a state of affairs prior to 1830 which would have led to Faraday's death in that year. Some of those would undoubtedly have led to faster technological progress, and some to slower. Which of them are we referring to in the '*if* . . . *then* . . .' statement? Which counts as 'other things being equal'? Try as we may, we shall not succeed in resolving this ambiguity within spacetime physics. There is no avoiding the fact that in spacetime exactly one thing happens in reality, and everything else is fantasy.

We are forced to conclude that, in spacetime physics, conditional statements whose premise is false ('if Faraday had died in 1830 . . .') have no meaning. Logicians call such statements *counter-factual conditionals*, and their status is a traditional paradox. We all know what such statements mean, yet as soon as we try to state their meaning clearly it seems to evaporate. The source of this paradox is not in logic or linguistics, it is in physics – in the false physics of spacetime. Physical reality is not a spacetime. It is a much bigger and more diverse entity, the *multiverse*. To a first approximation the multiverse is like a very large number of co-existing and slightly interacting spacetimes. If spacetime is like a stack of snapshots, each snapshot being the whole of space at one moment, then the multiverse is like a vast collection of such stacks. Even this (as we shall see) slightly inaccurate picture of the multiverse can already accommodate causes and effects. For in the multiverse there are almost certainly some universes in which Faraday died in 1830, and it is a matter of fact (not observable fact, but objective fact none the less) whether technological progress in those universes was or was not delayed relative to our own. There is nothing arbitrary about which variants of our universe the counter-factual 'if Faraday had died in 1830 . . .' refers to: it refers to *the variants which really occur* somewhere in the multiverse. That is what resolves the ambiguity. Appealing to imaginary universes does not work, because we can imagine any universes we like, in any proportions we like. But in the multiverse, universes are present in definite proportions, so it is meaningful to say that certain types of event are 'very rare' or 'very common' in the multiverse, and that some events follow others 'in most cases'. Most logically possible universes are not present at all – for example, there are no universes in which the charge on an electron is different from that in our universe, or in which the laws of quantum physics do not hold. The laws of physics that are implicitly referred to in the counter-factual are the laws that are actually obeyed in other universes, namely the laws of quantum theory. Therefore the '*if . . . then . . .*' statement can unambiguously be taken to mean 'in most universes in which Faraday died in 1830, technological progress was delayed relative to our own'. In general we may say that an event X causes an event Y in our universe if both X and Y occur in our universe, but in most variants of our universe in which X does not happen, Y does not happen either.

If the multiverse were literally a collection of spacetimes, the quantum concept of time would be the same as the classical one. As Figure 11.6 shows, time would still be a sequence of moments. The only difference would be that at a particular moment in the multiverse, many universes would exist instead of one. Physical reality at a particular moment would be, in effect, a 'super-snapshot' consisting of snapshots of many different versions of the whole of space. The whole of reality for the whole of time would be the stack of all the super-snapshots, just as classically it was a

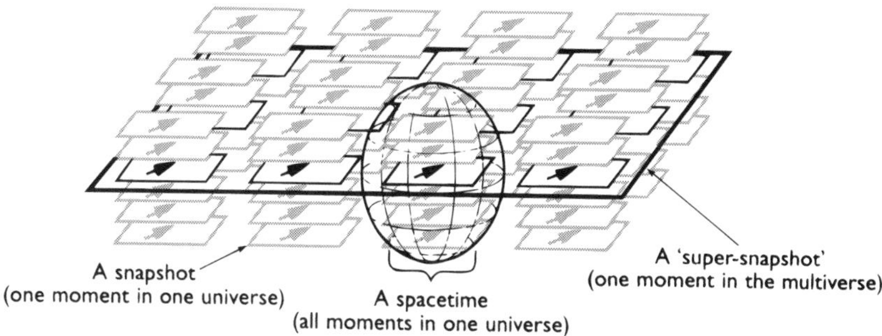

FIGURE 11.6 *If the multiverse were a collection of interacting spacetimes, time would still be a sequence of moments.*

stack of snapshots of space. Because of quantum interference, each snapshot would no longer be determined entirely by previous snapshots of the same spacetime (though it would approximately, because classical physics is often a good approximation to quantum physics). But the super-snapshots beginning with a particular moment would be entirely and exactly determined by the previous super-snapshots. This complete determinism would not give rise to complete predictability, even in principle, because making a prediction would require a knowledge of what had happened in all the universes, and each copy of us can directly perceive only one universe. Nevertheless, as far as the concept of time is concerned, the picture would be just like a spacetime with a sequence of moments related by deterministic laws, only with more happening at each moment, but most of it hidden from any one copy of any observer.

However, that is not quite how the multiverse is. A workable quantum theory of time – which would also be the quantum theory of gravity – has been a tantalizing and unattained goal of theoretical physics for some decades now. But we know enough about it to know that, though the laws of quantum physics are perfectly deterministic at the multiverse level, they do not partition the multiverse in the manner of Figure 11.6, into separate spacetimes, or into super-snapshots each of which entirely determines the others. So we know that the classical concept of time as a sequence of moments cannot be true, though it does provide a good approximation in many circumstances – that is, in many regions of the multiverse.

To elucidate the quantum concept of time, let us imagine that we have sliced the multiverse into a heap of individual snapshots, just as we did with spacetime. What can we glue them back together with? As before, the laws of physics and the intrinsic, physical properties of the snapshots are the only acceptable glue. If time in the multiverse were a sequence of moments, it would have to be possible to identify all the snapshots of space at a given moment, so as to make them into a super-snapshot. Not surprisingly, it turns out that there is no way of doing that. In the multiverse, snapshots do not have 'time stamps'. There is no such thing as which snapshot from another universe happens 'at the same moment' as a particular snapshot in our universe, for that would again imply that there is an overarching framework of time, outside the multiverse, relative to which events within the multiverse happen. There is no such framework.

Therefore there is no fundamental demarcation between snapshots of other times and snapshots of other universes. This is the distinctive core of the quantum concept of time:

Other times are just special cases of other universes.

This understanding first emerged from early research on quantum gravity in the 1960s, in particular from the work of Bryce DeWitt, but to the best of my knowledge it was not stated in a general way until 1983, by Don Page and William Wooters. The snapshots which we call 'other times in our universe' are distinguished from 'other universes' only from our perspective, and only in that they are especially closely related to ours by the laws of physics. They are therefore the ones of whose existence our own snapshot holds the most evidence. For that reason, we discovered them thousands of years before we discovered the rest of the multiverse, which impinges on us very weakly by comparison, through interference effects. We evolved special language constructs (past and future forms of verbs) for talking about them. We also evolved other constructs (such as '*if* . . . *then* . . .' statements, and conditional and subjunctive forms of verbs) for talking about other types of snapshot, without even knowing that they exist. We have traditionally placed these two types of snapshot – other times, and other universes – in entirely different conceptual categories. Now we see that this distinction is unnecessary.

Let us now proceed with our notional reconstruction of the multiverse. There are far more snapshots in our heap now, but let us again start with an individual snapshot of one universe at one moment. If we now search the heap for other snapshots that are very similar to the original one, we find that this heap is very different from the disassembled spacetime. For one thing, we find many snapshots that are absolutely identical to the original. In fact, any snapshot that is present at all is present in an infinity of copies. So it makes no sense to ask how many snapshots, numerically, have such-and-such a property, but only *what proportion* of the infinite total have that property. For the sake of brevity, when I speak of a certain 'number' of universes I shall always mean a certain proportion of the total number in the multiverse.

If, aside from *variants* of me in other universes, there are also multiple identical *copies* of me, which one am I? I am, of course, all of them. Each of them has just asked that question, 'which one am I?', and any true way of answering that question must give each of them the same answer. To assume that it is physically meaningful to ask which of the identical copies is me, is to assume that there is some frame of reference outside the multiverse, relative to which the answer could be given – 'I am the third one from the left . . .'. But what 'left' could that be, and what does 'the third one' mean? Such terminology makes sense only if we imagine the snapshots of me arrayed at different positions in some external space. But the multiverse does not exist in an external space any more than it exists in an external time: it contains all the space and time there is. It just exists, and physically it is all that exists.

Quantum theory does not in general determine what will happen in a particular snapshot, as spacetime physics does. Instead, it determines what proportion of all snapshots in the multiverse will have a given property. For this reason, we inhabitants of the multiverse can sometimes make only probabilistic predictions of our own experience, even though what will happen in the multiverse is completely determined. Suppose, for example, that we toss a coin. A typical prediction of quantum theory might be that *if*, in a certain number of snapshots, a coin has been set spinning in a certain manner and clocks show a certain reading, *then* there will also exist half that number of universes in which the clocks show a higher reading and the coin has fallen with 'heads' upwards, and another half in which the clocks show the higher reading and the coin has fallen with 'tails' upwards.

Figure 11.7 shows the small region of the multiverse in which these events happen. Even in that small region there are a lot of snapshots to illustrate, so we can spare only one point of the diagram for each snapshot. The snapshots we are looking at all contain clocks of some standard type, and the diagram is arranged so that all the snapshots with a particular clock reading appear in a vertical column, and the clock readings increase from left to right. As we scan along any vertical line in the diagram, not all the snapshots we pass through are different. We pass through groups of identical ones, as indicated by the shading. The snapshots in which clocks show the earliest reading are at the left edge of the diagram. We see that in all those snapshots, which are identical, the coin is spinning. At the right edge of the diagram, we see that in half the snapshots in which clocks show the latest reading the coin has fallen with 'heads' upwards, and in the other half it has fallen with 'tails' upwards. In universes with intermediate clock readings, three types of universe are present, in proportions that vary with the clock reading.

If you were present in the illustrated region of the multiverse, all copies of you would have seen the coin spinning at first. Later, half the copies of you would see 'heads' come up, and the other half would see 'tails'. At some intermediate stage you would have seen the coin in a state in which it is still in motion, but from which it is predictable which face it will show when it eventually settles down. This differentiation of identical copies of an observer

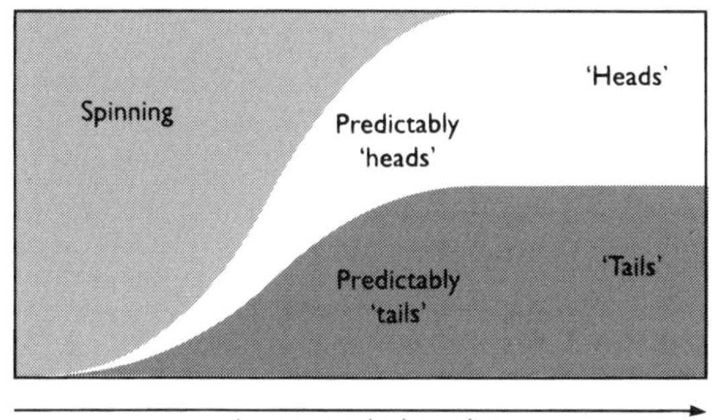

FIGURE 11.7 *A region of the multiverse containing a spinning coin. Each point in the diagram represents one snapshot.*

into slightly different versions is responsible for the subjectively probabilistic character of quantum predictions. For if you asked, initially, what result you were destined to see for the coin toss, the answer would be that that is strictly unpredictable, for half the copies of you that are asking that question would see 'heads' and the other half would see 'tails'. There is no such thing as 'which half' would see 'heads', any more than there is an answer to the question 'which one am I?'. For practical purposes you could regard this as a probabilistic prediction that the coin has a 50 per cent chance of coming up 'heads', and a 50 per cent chance of coming up 'tails'.

The determinism of quantum theory, just like that of classical physics, works both forwards and backwards in time. From the state of the combined collection of 'heads' and 'tails' snapshots at the later time in Figure 11.7, the 'spinning' state at an earlier time is completely determined, and vice versa. Nevertheless, from the point of view of any observer, information is lost in the coin-tossing process. For whereas the initial, 'spinning' state of the coin may be experienced by an observer, the final combined 'heads' and 'tails' state does not correspond to any possible experience of the observer. Therefore an observer at the earlier time may observe the coin and predict its future state, and the consequent subjective probabilities. But none of the later copies of the observer can possibly observe the information necessary to retrodict the 'spinning' state, for that information is by then distributed across two different types of universe, and that makes retrodiction from the final state of the coin impossible. For example, if all we know is that the coin is showing 'heads', the state a few seconds earlier might have been the state I called 'spinning', or the coin might have been spinning in the opposite direction, or it might have been showing 'heads' all the time. There is no possibility of retrodiction here, even probabilistic retrodiction. The earlier state of the coin is simply not determined by the later state of the 'heads' snapshots, but only by the joint state of the 'heads' and the 'tails' snapshots.

Any horizontal line across Figure 11.7 passes through a sequence of snapshots with increasing clock readings. We might be tempted to think of such a line – such as the one shown in Figure 11.8 – as a spacetime, and of the whole diagram as a stack of spacetimes, one for each such line. We can read off from Figure 11.8 what happens in the 'spacetime' defined by the horizontal line. For a period, it contains a spinning coin. Then, for a further period, it contains the coin moving in a way that will predictably result in 'heads'. But later, in contradiction to that, it contains the coin moving in a way that will predictably result in 'tails', and eventually it does show 'tails'. But this is merely a deficiency of the diagram, as I pointed out in Chapter 9 (see Figure 9.4, p. 212). In this case the laws of quantum mechanics predict that no observer who remembers seeing the coin in the 'predictably heads' state can see it in the 'tails' state: that is the justification for calling that state 'predictably heads' in the first place. Therefore no observer in the multiverse would recognize events as they occur in the 'spacetime'

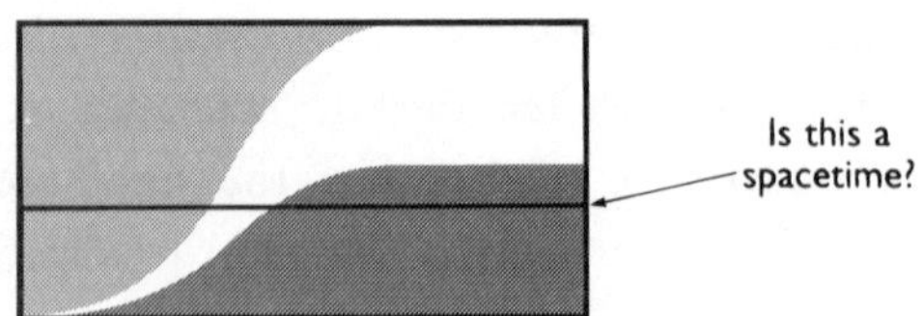

FIGURE 11.8 *A sequence of snapshots with increasing clock readings is not necessarily a spacetime.*

defined by the line. All this goes to confirm that we cannot glue the snapshots together in an arbitrary fashion, but only in a way that reflects the relationships between them that are determined by the laws of physics. The snapshots along the line in Figure 11.8 are not sufficiently interrelated to justify their being grouped together in a single universe. Admittedly they appear in order of increasing clock readings which, in *spacetime*, would be 'time stamps' which would be sufficient for the spacetime to be reassembled. But in the multiverse there are far too many snapshots for clock readings alone to locate a snapshot relative to the others. To do that, we need to consider the intricate detail of which snapshots determine which others.

In spacetime physics, any snapshot is determined by any other. As I have said, in the multiverse that is in general not so. Typically, the state of one group of identical snapshots (such as the ones in which the coin is 'spinning') determines the state of an equal number of differing snapshots (such as the 'heads' and 'tails' ones). Because of the time-reversibility property of the laws of quantum physics, the overall, multi-valued state of the latter *group* also determines the state of the former. However, in some regions of the multiverse, and in some places in space, the snapshots of some physical objects do fall, for a period, into chains, each of whose members determines all the others to a good approximation. Successive snapshots of the solar system would be the standard example. In such regions, classical physical laws are a good approximation to the quantum ones. In those regions and places, the multiverse does indeed look as in Figure 11.6, a collection of spacetimes, and at that level of approximation the quantum concept of time reduces to the classical one. One can distinguish approximately between 'different times' and 'different universes', and time is approximately a sequence of moments. But that approximation always breaks down if one examines the snapshots in more detail, or looks far forwards or backwards in time, or far afield in the multiverse.

All experimental results currently available to us are compatible with the approximation that time is a sequence of moments. We do not expect that approximation to break down in any foreseeable terrestrial experiment, but theory tells us that it must break down badly in certain types of physical process. The first is the beginning of the universe, the Big Bang. According to classical physics, time began at a moment when space was infinitely dense and occupied only a single point, and before that there were no moments. According to quantum physics (as best we can tell), the snapshots very near the Big Bang are not in any particular order. The sequential property of time does not begin at the Big Bang, but at some later time. In the nature of things, it does not make sense to ask how much later. But we can say that the earliest moments which are, to a good approximation, sequential occur roughly when classical physics would extrapolate that the Big Bang had happened 10^{-43} seconds (the *Planck time*) earlier.

A second and similar sort of breakdown of the sequence of time is thought to occur in the interiors of black holes, and at the final recollapse of the universe (the 'Big Crunch'), if there is one. In both cases matter is compressed to infinite density according to classical physics, just as at the Big Bang, and the resulting gravitational forces tear the fabric of spacetime apart.

By the way, if you have ever wondered what happened before the Big Bang, or what will happen after the Big Crunch, you can stop wondering now. Why is it hard to accept that there are no moments before the Big Bang or after the Big Crunch, so that nothing happens, or exists, there? Because it is hard to imagine time coming to a halt, or starting up. But then, time does not have to come to a halt or start up, for it does not move at all. The multiverse does not 'come into existence' or 'cease to exist'; those terms presuppose the flow of time. It is only imagining the flow of time that makes us wonder what happened 'before' or 'after' the whole of reality.

Thirdly, it is thought that on a sub-microscopic scale quantum effects again warp and tear the fabric of spacetime, and that closed loops of time – in effect, tiny time machines – exist on that scale. As we shall see in the next chapter, this sort of breakdown of the sequence of time is also physically possible on a large scale, and it is an open question whether it occurs near such objects as rotating black holes.

Thus, although we cannot yet detect any of these effects, our best theories already tell us that spacetime physics is never an exact description of reality. However good an approximation it is, time in reality must be fundamentally different from the linear sequence which common sense supposes. Nevertheless, everything in the multiverse is determined just as rigidly as in classical spacetime. Remove one snapshot, and the remaining ones determine it exactly. Remove *most* snapshots, and the few remaining ones may still determine everything that was removed, just as they do in spacetime. The difference is only that, unlike spacetime, the multiverse does not consist of the mutually determining layers I have called super-snapshots, which could serve as 'moments' of the multiverse. It is a complex, multi-dimensional jigsaw puzzle.

In this jigsaw-puzzle multiverse, which neither consists of a sequence of moments nor permits a flow of time, the common-sense concept of cause and effect makes perfect sense. The problem that we found with causation in spacetime was that it is a property of *variants* of the causes and effects, as well as of the causes and effects themselves. Since those variants existed only in our imagination, and not in spacetime, we ran up against the physical meaninglessness of drawing substantive conclusions from the imagined properties of non-existent ('counter-factual') physical processes. But in the multiverse variants do exist, in different proportions, and they obey definite, deterministic laws. Given these laws, it is an objective fact which events make a difference to the occurrence of which other events. Suppose that there is a group of snapshots, not necessarily identical, but all sharing the property X. Suppose that, given the existence of this group, the laws of physics determine that there exists another group of snapshots with property Y. One of the conditions for X to be a cause of Y has then been met. The other condition has to do with variants. Consider the variants of the first group that do not have the property X. If, from the existence of these, the existence of some of the Y snapshots is still determined, then X was not a cause of Y: for Y would have happened even without X. But if, from the group of non-X variants, only the existence of non-Y variants is determined, then X was a cause of Y.

There is nothing in this definition of cause and effect that logically requires causes to precede their effects, and it could be that in very exotic situations, such as very close to the Big Bang or inside black holes, they do not. In everyday experience, however, causes always precede their effects, and this is because – at least in our vicinity in the multiverse – the number of distinct types of snapshot tends to increase rapidly with time, and hardly ever decreases. This property is related to the second law of thermodynamics, which states that ordered energy, such as chemical or gravitational potential energy, may be converted entirely into disordered energy, i.e. heat, but never vice versa. Heat is microscopically random motion. In multiverse terms, this means many microscopically different states of motion in different universes. For example, in successive snapshots of the coin at ordinary magnifications, it seems that the setting-down process converts a group of identical 'predictably heads' snapshots into a group of identical 'heads' snapshots. But during that process the energy of the coin's motion is converted into heat, so at magnifications large enough to see individual molecules the latter group of snapshots are not identical at all. They all agree that the coin is in the 'heads' position, but they show its molecules, and those of the surrounding air and of the surface on which it lands, in many different configurations. Admittedly, the initial 'predictably heads' snapshots are not microscopically identical either, because some heat is present there too, but the production of heat in the process means that these snapshots are very much less diverse than the later ones. So each homogeneous group of 'predictably heads' snapshots determines the existence of – and therefore causes – vast numbers of microscopically different 'heads' snapshots. But no single 'heads' snapshot by itself determines the existence of any 'predictably heads' snapshots, and so is not a cause of them.

The conversion, relative to any observer, of possibilities into actualities – of an open future into a fixed past – also makes sense in this framework. Consider the coin-tossing example again. Before the coin toss, the future is open from the point of view of an observer, in the sense that it is still possible that either outcome, 'heads' or 'tails', will be observed by that observer. From that observer's point of view both outcomes are possibilities, even though objectively they are both actualities. After the coin has settled, the copies of the observer have differentiated into two groups. Each observer has observed, and remembers, only one outcome of the coin toss. Thus the outcome, once it is in the past of any observer, has become single-valued and actual for every copy of the observer, even though from the multiverse point of view it is just as two-valued as ever.

Let me sum up the elements of the quantum concept of time. Time is not a sequence of moments, nor does it flow. Yet our intuitions about the properties of time are broadly true. Certain events are indeed causes and effects of one another. Relative to an observer, the future is indeed open and the past fixed, and possibilities do indeed become actualities. The reason why our traditional theories of time are nonsense is that they try to express these true intuitions within the framework of a false classical physics. In quantum physics they make sense, because time was a quantum concept all along. We exist in multiple versions, in universes called 'moments'. Each version of us is not directly aware of the others, but has evidence of their existence because physical laws link the contents of different universes. It is tempting to suppose that the moment of which we are aware is the only real one, or is at least a little more real than the others. But that is just solipsism. All moments are physically real. The whole of the multiverse is physically real. Nothing else is.

TERMINOLOGY

flow of time The supposed motion of the present moment in the future direction, or the supposed motion of our consciousness from one moment to another. (This is nonsense!)

spacetime Space and time, considered together as a static four-dimensional entity.

spacetime physics Theories, such as relativity, in which reality is considered to be a spacetime. Because reality is a multiverse, such theories can at best be approximations.

free will The capacity to affect future events in any one of several possible ways, and to choose which shall occur.

counter-factual conditional A conditional statement whose premise is false (such as '*if* Faraday had died in 1830, *then* X would have happened').

snapshot (terminology for this chapter only) A universe at a particular time.

SUMMARY

Time does not flow. Other times are just special cases of other universes.

Time travel may or may not be feasible. But we already have a reasonably good theoretical understanding of what it would be like if it were, an understanding that involves all four strands.

12

Time Travel

It is a natural thought, given the idea that time is in some ways like an additional, fourth dimension of space, that if it is possible to travel from one place to another, perhaps it is also possible to travel from one time to another. We saw in the previous chapter that the idea of 'moving' through time, in the sense in which we move through space, does not make sense. Nevertheless, it seems clear what one would mean by travelling to the twenty-fifth century or to the age of the dinosaurs. In science fiction, time machines are usually envisaged as exotic vehicles. One sets the controls to the date and time of one's chosen destination, waits while the vehicle travels to that date and time (sometimes one can choose the place as well), and there one is. If one has chosen the distant future, one converses with conscious robots and marvels at interstellar spacecraft, or (depending on the political persuasion of the author) one wanders among charred, radioactive ruins. If one has chosen the distant past, one fights off an attack by a *Tyrannosaurus rex* while pterodactyls flutter overhead.

The presence of dinosaurs would be impressive evidence that we really had reached an earlier era. We should be able to cross-check this evidence, and determine the date more precisely, by observing some natural long-term 'calendar' such as the shapes of the constellations in the night sky or the relative proportions of various radioactive elements in rocks. Physics provides many such calendars, and the laws of physics cause them to agree with one another when suitably calibrated. According to the approximation that the multiverse consists of a set of parallel spacetimes, each consisting of a stack of 'snapshots' of space, the date defined in this way is a property of an entire snapshot, and any two snapshots are separated by a time interval which is the difference between their dates. Time travel is any process that causes a disparity between, on the one hand, this interval between two snapshots, and on the other, our own experience of how much time has elapsed between our being in those two snapshots. We might refer to a clock that we carry with us, or we might estimate how much thinking we have had the opportunity to do, or we might measure by physiological criteria how much our bodies have aged. If we observe that a long time has passed externally, while by all subjective measures we have experienced a much shorter time, then we have travelled into the future. If, on the other hand, we observe the external clocks and calendars indicating a particular time, and later (subjectively) we observe them consistently indicating an earlier time, then we have travelled into the past.

Most authors of science fiction realize that future- and past-directed time travel are radically different sorts of process. I shall not give future-directed time travel much attention here, because it is by far the less problematic proposition. Even in everyday life, for example when we sleep and wake up, our subjectively experienced time can be shorter than the external elapsed time. People who recover from comas lasting several years could be said to have travelled that many years into the future, were it not for the fact that their bodies have aged according to external time rather than the time they experienced subjectively. So, in principle, a technique similar to that which we envisaged in Chapter 5 for slowing down a virtual-reality user's brain could be applied to the whole body, and thus could be used for fully fledged future-directed time travel. A less intrusive method is provided by Einstein's special theory of relativity, which says that in general an observer who accelerates or decelerates experiences less time than an observer who is at rest or in uniform motion. For example, an astronaut who went on a round-trip involving acceleration to speeds close to that of light would experience much less time than an observer who remained on Earth. This effect is known as *time dilation*. By accelerating enough, one can make the duration of the flight from the astronaut's point of view as short as one likes, and the duration as measured on Earth as long as one likes. Thus one could travel as far into the future as one likes in a given, subjectively short time. But such a trip to the future is irreversible. The return journey would require past-directed time travel, and no amount of time dilation can allow a spaceship to return from a flight before it took off.

Virtual reality and time travel have this, at least, in common: they both systematically alter the usual relationship between external reality and the user's experience of it. So one might ask this question: if a universal virtual-reality generator could so easily be programmed to effect future-directed time travel, is there a way of using it for past-directed time travel? For instance, if slowing us down would send us into the future, would speeding us up send us into the past? No; the outside world would merely seem to slow down. Even at the unattainable limit where the brain operated infinitely fast, the outside world would appear frozen at a particular moment. That would still be time travel, by the above definition, but it would not be past-directed. One might call it 'present-directed' time travel. I remember wishing for a machine capable of present-directed time travel when doing last-minute revision for exams – what student has not?

Before I discuss past-directed time travel itself, what about the *rendering* of past-directed time travel? To what extent could a virtual-reality generator be programmed to give the user the experience of past-directed time travel? We shall see that the answer to this question, like all questions about the scope of virtual reality, tells us about physical reality as well.

The distinctive aspects of experiencing a past environment are, by definition, experiences of certain physical objects or processes – 'clocks' and 'calendars' – in states that occurred only at past times (that is, in past snapshots). A virtual-reality generator could, of course, render those objects in those states. For instance, it could give one the experience of living in the age of the dinosaurs, or in the trenches of the First World War, and it could make the constellations, dates on newspapers or whatever, appear correctly for those times. How correctly? Is there any fundamental limit on how accurately any given era could be rendered? The Turing principle says that a universal virtual-reality generator can be built, and could be programmed to render any physically possible environment, so clearly it could be programmed to render any environment that did once exist physically.

To render a time machine that had a certain repertoire of past destinations (and therefore also to render the destinations themselves), the program would have to include historical records of the environments at those destinations. In fact, it would need more than mere records, because the experience of time travel would involve more than merely seeing past events unfolding around one. Playing recordings of the past to the user would be mere image generation, not virtual reality. Since a real time traveller would participate in events and act back upon the past environment, an accurate virtual-reality rendering of a time machine, as of any environment, must be interactive. The program would have to calculate, for each action of the user, how the historical environment would have responded to that action. For example, to convince Dr Johnson that a purported time machine really had taken him to ancient Rome, we should have to allow him to do more than just watch passively and invisibly as Julius Caesar walked by. He would want to test the authenticity of his experiences by kicking the local rocks. He might kick Caesar – or at least, address him in Latin and expect him to reply in kind. What it means for a virtual-reality rendering of a time machine to be accurate is that the rendering should respond to such interactive tests in the same way as would the real time machine, and as would the real past environments to which it travelled. That should include, in this case, displaying a correctly behaving, Latin-speaking rendering of Julius Caesar.

Since Julius Caesar and ancient Rome were physical objects, they could, in principle, be rendered with arbitrary accuracy. The task differs only in degree from that of rendering the Centre Court at Wimbledon, including the spectators. Of course, the complexity of the requisite programs would be tremendous. More complex still, or perhaps even impossible in principle, would be the task of gathering the information required to write the programs to render specific human beings. But writing the programs is not the issue here. I am not asking whether we can find out enough about a past environment (or, indeed, about a present or future environment) to write a program that would render that environment specifically. I am asking whether the *set of all possible* programs for virtual-reality generators does or does not include one that gives a virtual-reality rendering of past-directed time travel and, if so, how accurate that rendering can be. If there were *no* programs rendering time travel, then the Turing principle would imply that time travel was physically impossible (because it says that everything that is physically possible can be rendered by some program). And on the face of it, there is indeed a problem here. Even though there are programs which accurately render past environments, there appear to be fundamental obstacles to using them to render time travel. These are the same obstacles that appear to prevent time travel itself, namely the so-called 'paradoxes' of time travel.

Here is a typical such paradox. I build a time machine and use it to travel back into the past. There I prevent my former self from building the time machine. But if the time machine is not built, I shall not be able to use it to travel into the past, nor therefore to prevent the time machine from being built. So do I make this trip or not? If I do, then I deprive myself of the time machine and therefore do not make the trip. If I do not make the trip, then I allow myself to build the time machine and so do make the trip. This is sometimes called the 'grandfather paradox', and stated in terms of using time travel to kill one's grandfather before he had any children. (And then, if he had no children, he could not have had any grandchildren, so who killed him?) These two forms of the paradox are the ones most commonly cited, and happen to require an element of violent conflict between the time traveller and people in the past, so one finds oneself wondering who will win. Perhaps the time traveller will be defeated, and the paradox avoided. But violence is not an essential part of the problem here. If I had a time machine, I could decide as follows: that if, today, my future self visits me, having set out from tomorrow, then tomorrow I *shall not* use my time machine; and that if I receive no such visitor today, then tomorrow I *shall* use the time machine to travel back to today and visit myself. It seems to follow from this decision that if I use the time machine then I shall not use it, and if I do not use it then I shall use it: a contradiction.

A contradiction indicates a faulty assumption, so such paradoxes have traditionally been taken as proofs that time travel is impossible. Another assumption that is sometimes challenged is that of free will – whether time travellers can choose in the usual way how to behave. One then concludes that if time machines did exist, people's free will would be impaired. They would somehow be unable to form intentions of the type I have described; or else, when they travelled in time, they would somehow systematically forget the resolutions they made before setting out. But it turns out that the faulty assumption behind the paradoxes is neither the existence of a time machine nor the ability of people to choose their actions in the usual way. All that is at fault is the classical theory of time, which I have already shown, for quite independent reasons, to be untenable.

If time travel really were logically impossible, a virtual-reality rendering of it would also be impossible. If it required a suspension of free will, then so would a virtual-reality rendering of it. The paradoxes of time travel can be expressed in virtual-reality terms as follows. The accuracy of a virtual-reality rendering is the faithfulness, as far as is perceptible, of the rendered environment to the intended one. In the case of time travel the intended environment is one that existed historically. But as soon as the rendered environment responds, as it is required to, to the user kicking it, it thereby becomes historically inaccurate because the real environment never did respond to the user: the user never did kick it. For example, the real Julius Caesar never met Dr Johnson. Consequently Dr Johnson, in the very act of testing the faithfulness of the rendering by conversing with Caesar, would destroy that faithfulness by creating a historically inaccurate Caesar. A rendering can *behave* accurately by being a faithful image of history, or it can *respond* accurately, but not both. Thus it would appear that, in one way or the other, a virtual-reality rendering of time travel is intrinsically incapable of being accurate – which is another way of saying that time travel could not be rendered in virtual reality.

But is this effect really an impediment to the accurate rendering of time travel? Normally, mimicking an environment's actual behaviour is not the aim of virtual reality: what counts is that it should respond accurately. As soon as you begin to play tennis on the rendered Wimbledon Centre Court, you make it behave differently from the way the real one is behaving. But that does not make the rendering any less accurate. On the contrary, that is what is required for accuracy. Accuracy, in virtual reality, means the closeness of the rendered behaviour to that which the original environment *would* exhibit if the user were present in it. Only at the beginning of the rendering does the rendered environment's state have to be faithful to the original. Thereafter it is not its state but its responses to the user's actions that have to be faithful. Why is that 'paradoxical' for renderings of time travel but not for other renderings – for instance, for renderings of ordinary travel?

It seems paradoxical because in renderings of past-directed time travel the user plays a unique double, or multiple, role. Because of the looping that is involved, where for instance one or more copies of the user may co-exist and interact, the virtual-reality generator is in effect required to *render the user* while simultaneously responding to the user's actions. For example, let us imagine that I am the user of a virtual-reality generator running a time-travel-rendering program. Suppose that when I switch on the program, the environment that I see around me is a futuristic laboratory. In the middle there is a revolving door, like those at the entrances of large buildings, except that this one is opaque and is almost entirely enclosed in an opaque cylinder. The only way in or out of the cylinder is a single entrance cut in its side. The door within revolves continuously. It seems at first sight that there is little one can do with this device except to enter it, go round one or more times with the revolving door, and come out again. But above the entrance is a sign: 'Pathway to the Past'. It is a time machine, a fictional,

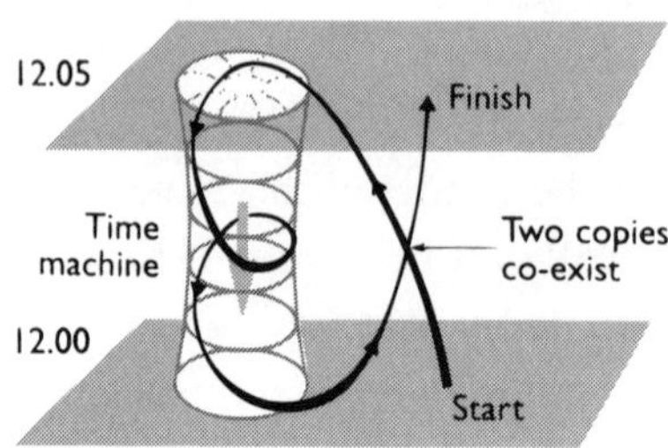

FIGURE 12.1 *Spacetime path taken by a time traveller.*

virtual-reality one. But if a real past-directed time machine existed it would, like this one, not be an exotic sort of vehicle but an exotic sort of *place*. Rather than drive or fly it to the past, one would take a certain path through it (perhaps using an ordinary space vehicle) and emerge at an earlier time.

On the wall of the simulated laboratory there is a clock, initially showing noon, and by the cylinder's entrance there are some instructions. By the time I have finished reading them it is five minutes past noon, both according to my own perception and according to the clock. The instructions say that if I enter the cylinder, go round once with the revolving door, and emerge, it will be five minutes earlier in the laboratory. I step into one of the compartments of the revolving door. As I walk round, my compartment closes behind me and then, moments later, reaches the entrance again. I step out. The laboratory looks much the same except – what? What exactly should I expect to experience next, if this is to be an accurate rendering of past-directed time travel?

Let me backtrack a little first. Suppose that by the entrance there is a switch whose two positions are labelled 'interaction *on*' and 'interaction *off*'. Initially it is at 'interaction *off*'. This setting does not allow the user to participate in the past, but only to observe it. In other words, it does not provide a full virtual-reality rendering of the past environment, but only image generation.

With this simpler setting at least, there is no ambiguity or paradox about what images ought to be generated when I emerge from the revolving door. They are images of me, in the laboratory, doing what I did at noon. One reason why there is no ambiguity is that I can remember those events, so I can test the images of the past against my own recollection of what happened. By restricting our analysis to a small, closed environment over a short period, we have avoided the problem analogous to that of finding out what Julius Caesar was really like, which is a problem about the ultimate limits of archaeology rather than about the inherent problems of time travel. In our case, the virtual-reality generator can easily obtain the information it needs to generate the required images, by making a recording of everything I do. Not, that is, a recording of what I do in physical reality (which is simply to lie still inside the virtual-reality generator), but of what I do in the virtual environment of the laboratory. Thus, the moment I emerge from the time machine, the virtual-reality generator ceases to render the laboratory at five minutes past noon, and starts to play back its recording, starting with images of what happened at noon. It displays this recording to me with the perspective adjusted for my present position and where I am looking, and it continuously readjusts the perspective in the usual way as I move. Thus, I see the clock showing noon again. I also see my earlier self, standing in front of the time machine, reading the sign above the entrance and studying the instructions, exactly as I did five minutes ago. I see him, but he cannot see me. No matter what I do, he – or rather *it*, the moving image of me – does not react to my presence in any way. After a while, it walks towards the time machine.

If I happen to be blocking the entrance, my image will nevertheless make straight for it and walk in, exactly as I did, for if it did anything else it would be an inaccurate image. There are many ways in which an image generator can be programmed to handle a situation where an image of a solid object has to pass through the user's location. For instance, the image could pass straight through like a ghost, or it could push the user irresistibly away. The latter option gives a more accurate rendering because then the images are to some extent tactile as well as visual. There need be no danger of my getting hurt as my image knocks me aside, however abruptly, because of course I am not physically there. If there is not enough room for me to get out of the way, the virtual-reality generator could make me flow effortlessly through a narrow gap, or even teleport me past an obstacle.

It is not only the image of myself on which I can have no further effect. Because we have temporarily switched from virtual reality to image generation, I can no longer affect anything in the simulated environment. If there is a glass of water on a table I can no longer pick it up and drink it, as I could have before I passed through the revolving door to the simulated past. By requesting a simulation of non-interactive, past-directed time travel, which is effectively a playback of specific events five minutes ago, I necessarily relinquish control over my environment. I cede control, as it were, to my former self.

As my image enters the revolving door, the time according to the clock has once again reached five minutes past twelve, though it is ten minutes into the simulation according to my subjective perception. What happens next depends on what I do. If I just stay in the laboratory, the virtual-reality generator's next task must be to place me at events that occur after five minutes past twelve, laboratory time. It does not yet have any recordings of such events, nor do I have any memories of them. Relative to me, relative to the simulated laboratory and relative to physical reality, those events have not yet happened, so the virtual-reality generator can resume its fully interactive rendering. The net effect is of my having spent five minutes in the past without being able to affect it, and then returning to the 'present' that I had left, that is, to the normal sequence of events which I can affect.

Alternatively, I can follow my image into the time machine, travel round the time machine with my image and emerge again into the laboratory's past. What happens then? Again, the clock says twelve noon. Now I can see *two* images of my former self. One of them is seeing the time machine for the first time, and notices neither me nor the other image. The second image appears to see the first but not me. I can see both of them. Only the first image appears to affect anything in the laboratory. This time, from the virtual-reality generator's point of view, nothing special has happened at the moment of time travel. It is still at the 'interaction *off*' setting, and is simply continuing to play back images of events five minutes earlier (from my subjective point of view), and these have now reached the moment when I began to see an image of myself.

After another five minutes have passed I can again choose whether to re-enter the time machine, this time in the company of *two* images of myself (Figure 12.2). If I repeat the process, then after every five subjective minutes an additional image of me will appear. Each image will appear to see all the ones that appeared earlier than itself (in my experience), but none of those that appeared later than itself.

If I continue the experience for as long as possible, the maximum number of copies of me that can co-exist will be limited only by the image generator's collision avoidance strategy. Let us assume that it tries to make it realistically difficult for me to squeeze myself into the revolving door with all my images. Then eventually I shall be forced to do something other than travel back to the past with them. I could wait a little, and take the compartment after theirs, in which case I should reach the laboratory a moment after they do. But that just postpones the problem of overcrowding in the time machine. If I keep going round this loop, eventually all the 'slots' for time travelling into the period of five minutes after noon will be filled, forcing me to let myself reach a later time from which there will be no further means of returning to that period. This too is a property that time machines would have if they existed physically. Not only are they places, they are places

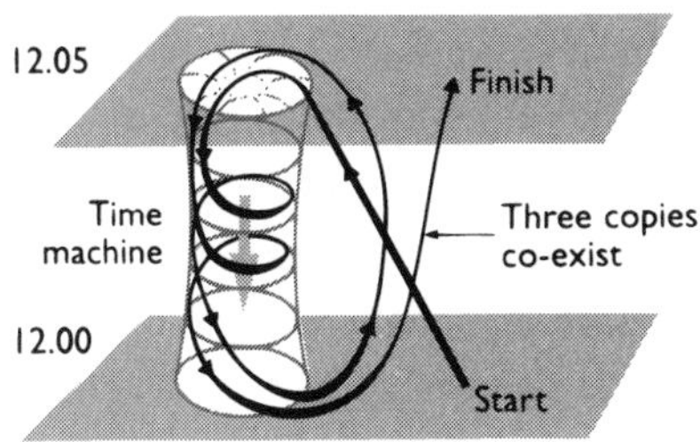

FIGURE 12.2 *Repeatedly using the time machine allows multiple copies of the time traveller to co-exist.*

with a finite capacity for supporting through traffic into the past.

Another consequence of the fact that time machines are not vehicles, but places or paths, is that one is not completely free to choose which time to use them to travel to. As this example shows, one can use a time machine only to travel to times and places at which it has existed. In particular, one cannot use it to travel back to a time before its construction was completed.

The virtual-reality generator now has recordings of many different versions of what happened in that laboratory between noon and five minutes past. Which one depicts the real history? We ought not be too concerned if there is no answer to this question, for it asks what is real in a situation where we have artificially suppressed interactivity, making Dr Johnson's test inapplicable. One could argue that only the last version, the one depicting the most copies of me, is the real one, because the previous versions all in effect show history from the point of view of people who, by the artificial rule of non-interaction, were prevented from fully seeing what was happening. Alternatively, one could argue that the first version of events, the one with a single copy of me, is the only real one because it is the only one I experienced interactively. The whole point of non-interactivity is that we are temporarily preventing ourselves from changing the past, and since subsequent versions all differ from the first one, they do not depict the past. All they depict is someone *viewing* the past by courtesy of a universal image generator.

One could also argue that all the versions are equally real. After all, when it is all over I remember having experienced not just one history of the laboratory during that five-minute period, but several such histories. I experienced them successively, but from the laboratory's point of view they all happened during the same five-minute period. The full record of my experience requires many snapshots of the laboratory for each clock-defined instant, instead of the usual single snapshot per instant. In other words, this was a rendering of parallel universes. It turns out that this last interpretation is the closest to the truth, as we can see by trying the same experiment again, this time with interaction switched on.

The first thing I want to say about the interactive mode, in which

I am free to affect the environment, is that *one* of the things I can choose to make happen is the exact sequence of events I have just described for the non-interactive mode. That is, I can go back and encounter one or more copies of myself, yet nevertheless (if I am a good enough actor) behave exactly as though I could not see some of them. Nevertheless, I must watch them carefully. If I want to recreate the sequence of events that occurred when I did this experiment with interaction switched off, I must remember what the copies of me do so that I can do it myself on subsequent visits to this time.

At the beginning of the session, when I first see the time machine, I immediately see it disgorging one or more copies of me. Why? Because with interaction switched on, when I come to use the time machine at five minutes past noon I shall have the right to affect the past to which I return, and that past is what is happening now, at noon. Thus my future self or selves are arriving to exercise their right to affect the laboratory at noon, and to affect me, and in particular to be seen by me.

The copies of me go about their business. Consider the computational task that the virtual-reality generator has to execute, in rendering these copies. There is now a new element that makes this overwhelmingly more difficult than it was in the non-interactive mode. How is the virtual-reality generator to find out what the copies of me are going to do? It does not yet have any recordings of that information, for in physical time the session has only just begun. Yet it must immediately present me with renderings of my future self.

So long as I am resolved to pretend that I cannot see these renderings, and then to mimic whatever I see them do, they are not going to be subjected to too stringent a test of accuracy. The virtual-reality generator need only make them do *something* – anything that I might do; or more precisely any behaviour that I am capable of mimicking. Given the technology that we are assuming the virtual-reality generator to be based on, that would presumably not be exceeding its capabilities. It has an accurate mathematical model of my body, and a degree of direct access to my brain. It can use these to calculate some behaviour which I could mimic, and then have its initial renderings of me carry out that behaviour.

So I begin the experience by seeing some copies of me emerge from the revolving door and do something. I pretend not to notice them, and after five minutes I go round the revolving door myself and mimic what I earlier saw the first copy doing. Five minutes later I go round again and mimic the second copy, and so on. Meanwhile, I notice that one of the copies always repeats what *I* had been doing during the first five minutes. At the end of the time-travelling sequence, the virtual-reality generator will again have several records of what happened during the five minutes after noon, but this time all those records will be identical. In other words, only one history happened, namely that I met my future self but pretended not to notice. Later I became that future self, travelled back in time to meet my past self, and was apparently not noticed. That is all very tidy and non-paradoxical – and unrealistic. It was achieved by the virtual-reality generator and me engaging in an intricate, mutually referential game: I was mimicking it while it was mimicking me. But with normal interactions switched on, I can choose not to play that game.

If I really had access to virtual-reality time travel, I should certainly want to test the authenticity of the rendering. In the case we are discussing, the testing would begin as soon as I saw the copies of me. Far from ignoring them, I would immediately engage them in conversation. I am far better equipped to test their authenticity than Dr Johnson would be to test Julius Caesar's. To pass even this initial test, the rendered versions of me would effectively have to be artificial intelligent beings – moreover, beings so similar to me, at least in their responses to external stimuli, that they can convince me they are accurate renderings of how I might be five minutes from now. The virtual-reality generator must be running programs similar in content and complexity to my mind. Once again, the difficulty of writing such programs is not the issue here: we are investigating the principle of virtual-reality time travel, not its practicality. It does not matter where our hypothetical virtual-reality generator gets its programs, for we are asking whether the *set of all possible programs* does or does not include one that accurately renders time travel. But our virtual-reality generator does in principle have the means of discovering all the possible ways I might behave in various situations. This information is located in the physical state of my brain, and sufficiently precise measurements could in principle read it out. One (probably unacceptable) method of doing this would be for the virtual-reality generator to cause my brain to interact, in virtual reality, with a test environment, record its behaviour and then restore its original state, perhaps by running it backwards. The reason why this is probably unacceptable is that I would presumably *experience* that test environment, and though I should not recall it afterwards, I want the virtual-reality generator to give me the experiences I specify and no others.

In any case, what matters for present purposes is that, since my brain is a physical object, the Turing principle says that it is within the repertoire of a universal virtual-reality generator. So it is possible in principle for the copy of me to pass the test of whether he accurately resembles me. But that is not the only test I want to perform. Mainly, I want to test whether the time travel itself is being rendered authentically. To that end I want to find out not just whether this person is authentically me, but whether he is authentically from the future. In part I can test this by questioning him. He should say that he remembers being in my position five minutes ago, and that he then travelled around the revolving door and met me. I should also find that *he* is testing *my* authenticity. Why would he do that? Because the most stringent and straightforward way in which I could test his resemblance to the future me would be to wait until I have passed through the time machine, and then look for two things: first, whether the copy of me whom I find there behaves as I remember myself behaving; and second, whether *I* behave as I remember the *copy* behaving.

In both these respects the rendering will certainly fail the test! At my very first and slightest attempt to behave differently from the way I remember my copy behaving, I shall succeed. And it will be almost as easy to make him behave differently from the way in which I behaved: all I have to do is ask him a question which I, in his place, had not been asked, and which has a distinctive answer. So however much they resemble me in appearance and personality, the people who emerge from the virtual-reality time machine are not authentic renderings of the person I am shortly to become. Nor should they be – after all, I have the firm intention not to behave as they do when it is my turn to use the time machine and, since the virtual-reality generator is now allowing me to interact freely with the rendered environment, there is nothing to prevent me from carrying out that intention.

Let me recap. As the experiment begins I meet a person who is recognizably me, apart from slight variations. Those variations consistently point to his being from the future: he remembers the laboratory at five minutes past noon, a time which, from my perspective, has not happened yet. He remembers setting out at that time, passing through the revolving door and arriving at noon. He remembers, before all that, beginning this experiment at noon and seeing the revolving door for the first time, and seeing copies of himself emerging. He says that this happened over five minutes ago, according to his subjective perception, though according to mine the whole experiment has not yet lasted five minutes. And so on. Yet though he passes all tests for being a version of me from the future, it is demonstrably not *my* future. When I test whether he is the specific person I am going to become, he fails that test. Similarly, he tells me that *I* fail the test for being his past self, since I am not doing exactly what he remembers himself doing.

So when I travel to the laboratory's past, I find that it is not the same past as I have just come from. Because of his interaction with me, the copy of me whom I find there does not behave quite as I remember behaving. Therefore, if the virtual-reality generator were to record the totality of what happens during this time-travel sequence, it would again have to store several snapshots for each instant as defined by the laboratory clock, and this time they would all be different. In other words, there would be several distinct, parallel histories of the laboratory during the five-minute time-travelling period. Again, I have experienced each of these histories in turn. But this time I have experienced them all interactively, so there is no excuse for saying that any of them are less real than the others. So what is being rendered here is a little multiverse. If this were physical time travel, the multiple snapshots at each instant would be parallel universes. Given the quantum concept of time, we should not be surprised at this. We know that the snapshots which stack themselves approximately into a single time sequence in our everyday experience are in fact parallel universes. We do not normally experience the other parallel universes that exist at the *same* time, but we have reason to believe that they are there. So, if we find some method, as yet unspecified, of travelling to an earlier time, why should we expect that method necessarily to take each copy of us to the particular snapshot which that copy had already experienced? Why should we expect every visitor we receive from the future to hail from the particular future snapshots in which we shall eventually find ourselves? We really should not expect this. Asking to be allowed to interact with the past environment means asking to change it, which means by definition asking to be in a different snapshot of it from the one we remember. A time traveller would return to the same snapshot (or, what is perhaps the same thing, to an identical snapshot) only in the extremely contrived case I discussed above, where no effective interaction takes place between the copies who meet, and the time traveller manages to make all the parallel histories identical.

Now let me subject the virtual-reality time machine to the ultimate test. Let me deliberately set out to enact a paradox. I form the firm intention that I stated above: I resolve that if a copy of me emerges at noon from the time machine, then I *shall not* enter it at five minutes past noon, or indeed at any time during the experiment. But if no one emerges, then at five minutes past noon I *shall* enter the time machine, emerge at noon, and then not use the time machine again. What happens? Will someone emerge from the time machine or not? Yes. And no! It depends which universe we are talking about. Remember that more than one thing happens in that laboratory at noon. Suppose that I see no one emerging from the time machine, as illustrated at the point marked 'Start' at the right of Figure 12.3. Then, acting on my firm intention, I wait until five minutes past noon and then walk round that now-familiar revolving door. Emerging at noon, I find, of course, another version of myself, standing at the point marked 'Start' on the *left* of Figure 12.3. As we converse, we find that he and I had formed the same intention. Therefore, because I have emerged into his universe, he will behave differently from the way I behaved. Acting on the same intention as mine leads him *not* to use the time machine. From then on, he and I can continue to interact for as long as the simulation lasts, and there will be two versions of me in that universe. In the universe I came from, the laboratory remains empty after five minutes past twelve, for I never return to it. We have encountered no paradox. Both versions of me have succeeded in enacting our shared intention – which was therefore not, after all, logically incapable of being carried out.

I and my alter ego in this experiment have had different experiences. He saw someone emerging from the time machine at noon, and I did not. Our experiences would have been equally faithful to our intention, and equally non-paradoxical, had our roles been reversed. That is, I could have seen him emerging from the time machine at noon, and then not used it myself. In that case both of us would have ended up in the universe I started in. In the universe he started in, the laboratory would remain empty.

Which of these two self-consistent possibilities will the virtual-

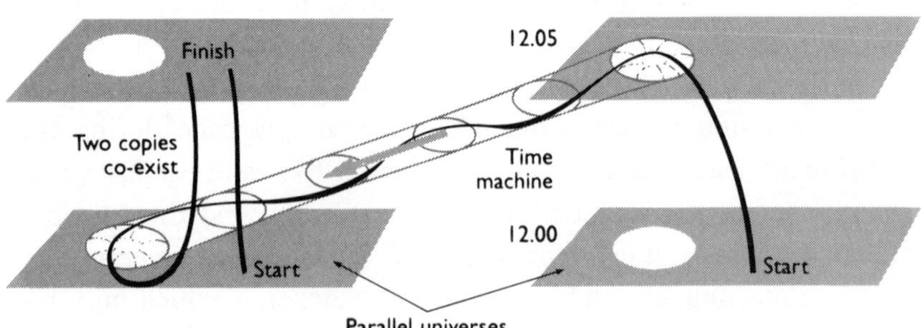

FIGURE 12.3 *Multiverse paths of a time traveller trying to 'enact a paradox'.*

reality generator show me? During this rendering of an intrinsically multiversal process, I play only one of the two copies of me; the program renders the other copy. At the beginning of the experiment the two copies look identical (though in physical reality they are different because only one of them is connected to a physical brain and body outside the virtual environment). But in the physical version of the experiment – if a time machine existed physically – the two universes containing the copies of me who were going to meet would initially be strictly identical, and both copies would be equally real. At the multiverse-moment when we met (in one universe) or did not meet (in the other), those two copies would become different. It is not meaningful to ask *which* copy of me would have which experience: so long as we are identical, there is no such thing as 'which' of us. Parallel universes do not have hidden serial numbers: they are distinguished only by what happens in them. Therefore in rendering all this for the benefit of one copy of me, the virtual-reality generator must recreate for me the effect of existing as two identical copies who then become different and have different experiences. It can cause that literally to happen by choosing at random, with equal probabilities, which of the two roles it will play (and therefore, given my intention, which role I shall play). For choosing randomly means in effect tossing some electronic version of a fair coin, and a fair coin is one that shows 'heads' in half the universes in which it is tossed and 'tails' in the other half. So in half the universes I shall play one role, and in the other half, the other. That is exactly what would happen with a real time machine.

We have seen that a virtual-reality generator's ability to render time travel accurately depends on its having detailed information about the user's state of mind. This may make one briefly wonder whether the paradoxes have been genuinely avoided. If the virtual-reality generator knows what I am going to do in advance, am I really free to perform whatever tests I choose? We need not get into any deep questions about the nature of free will here. I am indeed free to do whatever I like in this experiment, in the sense that for every possible way I may choose to react to the simulated past – including randomly, if I want to – the virtual-reality generator allows me to react in that way. And all the environments I interact with are affected by what I do, and react back on me in precisely the way they would if time travel were not taking place.

The reason why the virtual-reality generator needs information from my brain is not to predict *my* actions, but to render the behaviour of my counterparts from other universes. Its problem is that in the real version of this situation there would be parallel-universe counterparts of me, initially identical and therefore possessing the same propensities as me and making the same decisions. (Farther away in the multiverse there would also be others who were already different from me at the outset of the experiment, but a time machine would never cause me to meet those versions.) If there were some other way of rendering these people, the virtual-reality generator would not need any information from my brain, nor would it need the prodigious computational resources that we have been envisaging. For example, if some people who know me well were able to mimic me to some degree of accuracy (apart from external attributes such as appearance and tone of voice, which are relatively trivial to render) then the virtual-reality generator could use those people to act out the roles of my parallel-universe counterparts, and could thereby render time travel to that same degree of accuracy.

A real time machine, of course, would not face these problems. It would simply provide pathways along which I and my counterparts, who already existed, could meet, and it would constrain neither our behaviour nor our interactions when we did meet. The ways in which the pathways interconnect – that is, which snapshots the time machine would lead to – would be affected by my physical state, including my state of mind. That is no different from the usual situation, in which my physical state, as reflected in my propensity to behave in various ways, affects what happens. The great difference between this and everyday experience is that each copy of me is potentially having a large effect on other universes (by travelling to them).

Does being able to travel to the past of other universes, but not our own, really amount to time travel? Is it just *inter-universe* travel that makes sense, rather than time travel? No. The processes I have been describing really are time travel. First of all, it is not the case that we *cannot* travel to a snapshot where we have already been. If we arrange things correctly, we can. Of course if we change anything in the past – if we make it different from how it was in the past we came from – then we find ourselves in a different past. Fully fledged time travel would allow us to change the past. In other words, it allows us to make the past different from the way we remember it (in this universe). That means different from the way it actually is, in the snapshots in which we did not arrive to change anything. And those include, by definition, the snapshots we remember being in.

So wanting to change the specific past snapshots in which we once were does indeed not make sense. But that has nothing to do with time travel. It is a nonsense that stems directly from the nonsensical classical theory of the flow of time. Changing the past means choosing which past snapshot to be in, not changing any specific past snapshot into another one. In this respect, changing the past is no different from changing the future, which we do all the time. Whenever we make a choice, we change the future: we change it from what it would have been had we chosen differently. Such an idea would make no sense in classical, spacetime physics with its single future determined by the present. But it does make sense in quantum physics. When we make a choice, we change the future from what it will be in universes in which we choose differently. But in no case does any particular snapshot in the future change. It cannot change, for there is no flow of time with respect to which it could change. 'Changing' the future means choosing which snapshot we will be in; 'changing' the past means exactly the same thing. Because there is no flow of time, there is no such thing as changing a particular past snapshot, such as one we remember being in. Nevertheless, if we somehow gain physical access to the past, there is no reason why we could not change it in precisely the sense in which we change the future, namely by choosing to be in a different snapshot from the one we would have been in if we had chosen differently.

Arguments from virtual reality help in understanding time travel because the concept of virtual reality requires one to take 'counter-factual events' seriously, and therefore the multi-universe quantum concept of time seems natural when it is rendered in virtual reality. By seeing that past-directed time travel is within the repertoire of a universal virtual-reality generator, we learn that the idea of past-directed time travel makes perfect sense. But that is not to say that it is necessarily physically achievable. After all, faster-than-light travel, perpetual motion machines and many other physical impossibilities are all possible in virtual reality. No amount of reasoning about virtual reality can prove that a given process is permitted by the laws of physics (though it can prove that it is not: if we had reached the contrary conclusion, it would have implied, via the Turing principle, that time travel cannot occur physically). So what do our positive conclusions about virtual-reality time travel tell us about physics?

They tell us what time travel would look like if it did occur. They tell us that past-directed time travel would inevitably be a process set in several interacting and interconnected universes. In that process, the participants would in general travel from one universe to another whenever they travelled in time. The precise ways in which the universes were connected would depend, among other things, on the participants' states of mind.

So, for time travel to be physically possible it is necessary for there to be a multiverse. And it is necessary that the physical laws governing the multiverse be such that, in the presence of a time machine and potential time travellers, the universes become inter-connected in the way I have described, and not in any other way. For example, if I am not going to use a time machine come what may, then no time-travelling versions of me must appear in my snapshot; that is, no universes in which versions of me do use a time machine can become connected to my universe. If I am definitely going to use the time machine, then my universe must become connected to another universe in which I also definitely use it. And if I am going to try to enact a 'paradox' then, as we have seen, my universe must become connected with another one in which a copy of me has the same intention as I do, but by carrying out that intention ends up behaving differently from me. Remarkably, all this is precisely what quantum theory does predict. In short, the result is that if pathways into the past do exist, travellers on them are free to interact with their environment in just the same way as they could if the pathways did not lead into the past. In no case does time travel become inconsistent, or impose special constraints on time travellers' behaviour.

That leaves us with the question whether it is physically possible for pathways into the past to exist. This question has been the subject of much research, and is still highly controversial. The usual starting-point is a set of equations which form the (predictive) basis of Einstein's general theory of relativity, currently our best theory of space and time. These equations, known as *Einstein's equations*, have many solutions, each describing a possible four-dimensional configuration of space, time and gravity. Einstein's equations certainly permit the existence of pathways into the past; many solutions with that property have been discovered. Until recently, the accepted practice has been systematically to ignore such solutions. But this has not been for any reason arising from within the theory, nor from any argument within physics at all. It has been because physicists were under the impression that time travel would 'lead to paradoxes', and that such solutions of Einstein's equations must therefore be 'unphysical'. This arbitrary second-guessing is reminiscent of what happened in the early years of general relativity, when the solutions describing the Big Bang and an expanding universe were rejected by Einstein himself. He tried to change the equations so that they would describe a static universe instead. Later he referred to this as the biggest mistake of his life, and the expansion was verified experimentally by the American astronomer Edwin Hubble. For many years also, the solutions obtained by the German astronomer Karl Schwarzschild, which were the first to describe black holes, were mistakenly rejected as 'unphysical'. They described counter-intuitive phenomena, such as a region from which it is in principle impossible to escape, and gravitational forces becoming infinite at the black hole's centre. The prevailing view nowadays is that black holes do exist, and do have the properties predicted by Einstein's equations.

Taken literally, Einstein's equations predict that travel into the past would be possible in the vicinity of massive, spinning objects, such as black holes, if they spun fast enough, and in certain other situations. But many physicists doubt that these predictions are realistic. No sufficiently rapidly spinning black holes are known, and it has been argued (inconclusively) that it may be impossible to spin one up artificially, because any rapidly spinning material that one fired in might be thrown off and be unable to enter the black hole. The sceptics may be right, but in so far as their reluctance to accept the possibility of time travel is rooted in a belief that it leads to paradoxes, it is unjustified.

Even when Einstein's equations have been more fully understood, they will not provide conclusive answers on the subject of time travel. The general theory of relativity predates quantum theory and is not wholly compatible with it. No one has yet succeeded in formulating a satisfactory quantum version – a quantum theory of gravity. Yet, from the arguments I have given, quantum effects would be dominant in time-travelling situations. Typical candidate versions of a quantum theory of gravity not only allow past-directed connections to exist in the multiverse, they predict that such connections are continually forming and breaking spontaneously. This is happening throughout space and time, but only on a submicroscopic scale. The typical pathway formed by these effects is about 10^{-35} metres across, remains open for one Planck time (about 10^{-43} seconds), and therefore reaches only about one Planck time into the past.

Future-directed time travel, which essentially requires only efficient rockets, is on the moderately distant but confidently foreseeable technological horizon. Past-directed time travel, which requires the manipulation of black holes, or some similarly violent gravitational disruption of the fabric of space and time, will be practicable only in the remote future, if at all. At present we know of nothing in the laws of physics that rules out past-directed time travel; on the contrary, they make it plausible that time travel is possible. Future discoveries in fundamental physics may change this. It may be discovered that quantum fluctuations in space and time become overwhelmingly strong near time machines, and effectively seal off their entrances (Stephen Hawking, for one, has argued that some calculations of his make this likely, but his argument is inconclusive). Or some hitherto unknown phenomenon may rule out past-directed time travel – or provide a new and easier method of achieving it. One cannot predict the future growth of knowledge. But if the future development of fundamental physics continues to allow time travel in principle, then its practical attainment will surely become a mere technological problem that will eventually be solved.

Because no time machine provides pathways to times earlier than the moment at which it came into existence, and because of the way in which quantum theory says that universes are interconnected, there are some limits to what we can expect to learn by using time machines. Once we have built one, but not before, we may expect visitors, or at least messages, from the future to emerge from it. What will they tell us? One thing they will certainly not tell us is news of our own future. The deterministic nightmare of the prophecy of an inescapable future doom, brought about in spite of – or perhaps as the very consequence of – our attempts to avoid it, is the stuff of myth and science fiction only. Visitors from the future cannot know our future any more than we can, for they did not come from there. But they can tell us about the future of their universe, whose past was identical to ours. They can bring taped news and current affairs programmes, and newspapers with dates starting from tomorrow and onwards. If their society made some mistaken decision, which led to disaster, they can warn us of it. We may or may not follow their advice. If we follow it, we may avoid the disaster, or – there can be no guarantees – we may find that the result is even worse than what happened to them.

On average, though, we should presumably benefit greatly from studying their future history. Although it is not our future history, and although knowing of a possible impending disaster is not the same thing as knowing what to do about it, we should presumably learn much from such a detailed record of what, from our point of view, *might* happen.

Our visitors might bring details of great scientific and artistic achievements. If these were made in the near future of the other universe, it is likely that counterparts of the people who made them would exist in our universe, and might already be working towards those achievements. All at once, they would be presented with completed versions of their work. Would they be grateful? There is another apparent time-travel paradox here. Since it does not appear to create inconsistencies, but merely curiosities, it has been discussed more in fiction than in scientific arguments against time travel (though some philosophers, such as Michael Dummett, have taken it seriously). I call it the *knowledge paradox* of time travel; here is how the story typically goes. A future historian with an interest in Shakespeare uses a time machine to visit the great playwright at a time when he is writing *Hamlet*. They have a conversation, in the course of which the time traveller shows Shakespeare the text of Hamlet's 'To be or not to be' soliloquy, which he has brought with him from the future. Shakespeare likes it and incorporates it into the play. In another version, Shakespeare dies and the time traveller assumes his identity, achieving success by pretending to write plays which he is secretly copying from the *Complete Works of Shakespeare*, which he brought with him from the future. In yet another version, the time traveller is puzzled by not being able to locate Shakespeare at all. Through some chain of accidents, he finds himself impersonating Shakespeare and, again, plagiarizing his plays. He likes the life, and years later he realizes that he has become *the* Shakespeare: there never had been another one.

Incidentally, the time machine in these stories would have to be provided by some extraterrestrial civilization which had already achieved time travel by Shakespeare's day, and which was willing to allow our historian to use one of their scarce, non-renewable slots for travelling back to that time. Or perhaps (even less likely,

I guess) there might be a usable, naturally occurring time machine in the vicinity of some black hole.

All these stories relate a perfectly consistent chain – or rather, circle – of events. The reason why they are puzzling, and deserve to be called paradoxes, lies elsewhere. It is that in each story great literature comes into existence without anyone having written it: no one originally wrote it, no one has created it. And that proposition, though logically consistent, profoundly contradicts our understanding of where knowledge comes from. According to the epistemological principles I set out in Chapter 3, knowledge does not come into existence fully formed. It exists only as the result of creative processes, which are step-by-step, evolutionary processes, always starting with a problem and proceeding with tentative new theories, criticism and the elimination of errors to a new and preferable problem-situation. This is how Shakespeare wrote his plays. It is how Einstein discovered his field equations. It is how all of us succeed in solving any problem, large or small, in our lives, or in creating anything of value.

It is also how new living species come into existence. The analogue of a 'problem' in this case is an ecological niche. The 'theories' are genes, and the tentative new theories are mutated genes. The 'criticism' and 'elimination of errors' are natural selection. Knowledge is created by intentional human action, biological adaptations by a blind, mindless mechanism. The words we use to describe the two processes are different, and the processes are physically dissimilar too, but the detailed laws of epistemology that govern them both are the same. In one case they are called Popper's theory of the growth of scientific knowledge; in the other, Darwin's theory of evolution. One could formulate a knowledge paradox just as well in terms of living species. Say we take some mammals in a time machine to the age of the dinosaurs, when no mammals had yet evolved. We release our mammals. The dinosaurs die out and our mammals take over. Thus new species have come into existence without having evolved. It is even easier to see why this version is philosophically unacceptable: it implies a non-Darwinian origin of species, and specifically *creationism*. Admittedly, no Creator in the traditional sense is invoked. Nevertheless, the origin of species in this story is distinctly supernatural: the story gives no explanation – and *rules out the possibility of there being an explanation* – of how the specific and complex adaptations of the species to their niches got there.

In this way, knowledge-paradox situations violate epistemological or, if you like, evolutionary principles. They are paradoxical only because they involve the creation, out of nothing, of complex human knowledge or of complex biological adaptations. Analogous stories with other sorts of object or information on the loop are not paradoxical. Observe a pebble on a beach; then travel back to yesterday, locate the pebble elsewhere and move it to where you are going to find it. Why did you find it at that particular location? Because you moved it there. Why did you move it there? Because you found it there. You have caused some information (the position of the pebble) to come into existence on a self-consistent loop. But so what? The pebble had to be somewhere. Provided the story does not involve getting something for nothing, by way of knowledge or adaptation, it is no paradox.

In the multiverse view, the time traveller who visits Shakespeare has not come from the future of that copy of Shakespeare. He can affect, or perhaps replace, the copy he visits. But he can never visit the copy who existed in the universe he started from. And it is *that* copy who wrote the plays. So the plays had a genuine author, and there are no paradoxical loops of the kind envisaged in the story. Knowledge and adaptation are, even in the presence of pathways to the past, brought into existence only incrementally, by acts of human creativity or biological evolution, and in no other way.

I wish I could report that this requirement is also rigorously implemented by the laws that quantum theory imposes on the multiverse. I expect it is, but this is hard to prove because it is hard to express the desired property in the current language of theoretical physics. What mathematical formula distinguishes 'knowledge' or 'adaptation' from worthless information? What physical attributes distinguish a 'creative' process from a non-creative one? Although we cannot yet answer these questions, I do not think that the situation is hopeless. Remember the conclusions of Chapter 8, about the significance of life, and of knowledge, in the multiverse. I pointed out there (for reasons quite unconnected with time travel) that knowledge creation and biological evolution are physically significant processes. And one of the reasons was that those processes, and only those, have a particular effect on parallel universes – namely to create trans-universe structure by making them become alike. When, one day, we understand the details of this effect, we may be able to define knowledge, adaptation, creativity and evolution in terms of the convergence of universes.

When I 'enact a paradox', there are eventually two copies of me in one universe and none in the other. It is a general rule that after time travel has taken place the total number of copies of me, counted across all universes, is unchanged. Similarly, the usual conservation laws for mass, energy and other physical quantities continue to hold for the multiverse as a whole, though not necessarily in any one universe. However, there is no conservation law for knowledge. Possession of a time machine would allow us access to knowledge from an entirely new source, namely the creativity of minds in other universes. They could also receive knowledge from us, so one can loosely speak of a 'trade' in knowledge – and indeed a trade in artefacts embodying knowledge – across many universes. But one cannot take that analogy too literally. The multiverse will never be a free-trade area because the laws of quantum mechanics impose drastic restrictions on which snapshots can be connected to which others. For one thing, two universes first become connected only at a moment when they are identical: becoming connected makes them begin to diverge. It is only when those differences have accumulated, and new knowledge has been created in one universe and sent back in time to the other, that we could receive knowledge that does not already exist in our universe.

A more accurate way of thinking about the inter-universe 'trade' in knowledge is to think of all our knowledge-generating processes, our whole culture and civilization, and all the thought processes in the minds of every individual, and indeed the entire evolving biosphere as well, as being a gigantic *computation*. The whole thing is executing a self-motivated, self-generating computer program. More specifically it is, as I have mentioned, a virtual-reality program in the process of rendering, with ever-increasing accuracy, the whole of existence. In other universes there are other versions of this virtual-reality generator, some identical, some very different. If such a virtual-reality generator had access to a time machine, it would be able to receive some of the results of computations performed by its counterparts in other universes, in so far as the laws of physics allowed the requisite interchange of information. Each piece of knowledge that one obtains from a time machine will have had an author somewhere in the multiverse, but it may benefit untold numbers of different universes. So a time machine is a computational resource that allows certain types of computation to be performed with enormously greater efficiency than they could be on any individual computer. It achieves this efficiency by effectively sharing computational work among copies of itself in different universes.

In the absence of time machines, there tends to be very little interchange of information between universes because the laws of physics predict, in that case, very little causal contact between them. To a good degree of approximation, knowledge created in one set of identical snapshots reaches relatively few other snapshots, namely those that are stacked into spacetimes to the future of the original snapshots. But this is only an approximation. Interference phenomena are the result of causal contact between nearby universes. We have seen in Chapter 9 that even this minuscule level of contact can be used to exchange significant, computationally useful information between universes.

The study of time travel provides an arena – albeit at present only a theoretical, thought-experiment arena – in which we can see writ large some of the connections between what I call the 'four main strands'. All four strands play essential roles in the explanation of time travel. Time travel may be achieved one day, or it may not. But if it is, it should not require any fundamental change in world-view, at least for those who broadly share the world-view I am presenting in this book. All the connections that it could set up between past and future are comprehensible and non-paradoxical. And all the connections that it would necessitate, between apparently unconnected fields of knowledge, are there anyway.

TERMINOLOGY

time travel It is only past-directed time travel that really deserves the name.

past-directed In past-directed time travel the traveller experiences the same instant, as defined by external clocks and calendars, more than once in subjective succession.

future-directed In future-directed time travel the traveller reaches a later instant in a shorter subjective time than that defined by external clocks and calendars.

time machine A physical object that enables the user to travel into the past. It is better thought of as a place, or pathway, than as a vehicle.

paradox of time travel An apparently impossible situation that a time traveller could bring about if time travel were possible.

grandfather paradox A paradox in which one travels into the past and then prevents oneself from ever doing so.

knowledge paradox A paradox in which knowledge is created from nothing, through time travel.

SUMMARY

Time travel may or may not be achieved one day, but it is not paradoxical. If one travels into the past one retains one's normal freedom of action, but in general ends up in the past of a different universe. The study of time travel is an area of theoretical study in which all four of my main strands are significant: quantum mechanics, with its parallel universes and the quantum concept of time; the theory of computation, because of the connections between virtual reality and time travel, and because the distinctive features of time travel can be analysed as new modes of computation; and epistemology and the theory of evolution, because of the constraints they impose on how knowledge can come into existence.

Not only are the four strands related as part of the fabric of reality, there are also remarkable parallels between the four fields of knowledge as such. All four basic theories have the unusual status of being simultaneously accepted and rejected, relied upon and disbelieved, by most people working in those fields.

13

The Four Strands

A widely held stereotype of the scientific process is that of the idealistic young innovator pitted against the old fogies of the scientific 'establishment'. The fogies, hidebound by the comfortable orthodoxy of which they have made themselves both defenders and prisoners, are enraged by any challenge to it. They behave irrationally. They refuse to listen to criticism, engage in argument or accept evidence, and they try to suppress the innovator's ideas.

This stereotype has been elevated into a philosophy by Thomas Kuhn, author of the influential book *The Structure of Scientific Revolutions*. According to Kuhn, the scientific establishment is defined by its members' belief in the set of prevailing theories, which together form a world-view, or *paradigm*. A paradigm is the psychological and theoretical apparatus through which its holders observe and explain everything in their experience. (Within any reasonably self-contained area of knowledge, such as physics, one may also speak of the 'paradigm' within that field.) Should any observation seem to violate the relevant paradigm, its holders are simply blind to the violation. When confronted with evidence of it, they are obliged to regard it as an 'anomaly', an experimental error, a fraud – anything at all that will allow them to hold the paradigm inviolate. Thus Kuhn believes that the scientific values of openness to criticism and tentativeness in accepting theories, and the scientific methods of experimental testing and the abandonment of prevailing theories when they are refuted, are largely myths that it would not be humanly possible to enact when dealing with any significant scientific issue.

Kuhn accepts that, for *insignificant* scientific issues, something like a scientific process (as I outlined in Chapter 3) does happen. For he believes that science proceeds in alternating eras: there is 'normal science' and there is 'revolutionary science'. During an era of normal science nearly all scientists believe in the prevailing fundamental theories, and try hard to fit all their observations and subsidiary theories into that paradigm. Their research consists of tying up loose ends, of improving the practical applications of theories, of classifying, reformulating and confirming. Where applicable, they may well use methods that are scientific in the Popperian sense, but they never discover anything fundamental because they never question anything fundamental. Then along come a few young troublemakers who deny some fundamental tenet of the existing paradigm. This is not really scientific criticism, for the troublemakers are not amenable to reason either. It is just that they view the world through a new and different paradigm. How did they come by this paradigm? The pressure of accumulated evidence, and the inelegance of explaining it away under the old paradigm, finally got through to them. (Fair enough, though it is hard to see how one could succumb to pressure in the form of evidence to which one is, by hypothesis, blind.) Anyway, an era of 'revolutionary' science begins. The majority, who are still trying to do 'normal' science in the old paradigm, fight back by fair means and foul – interfering with publication, excluding the heretics from academic posts, and so on. The heretics manage to find ways of publishing, they ridicule the old fogies and they try to infiltrate influential institutions. The explanatory power of the new paradigm, in its own terms (for in terms of the old paradigm its explanations seem extravagant and unconvincing), attracts recruits from the ranks of uncommitted young scientists. There may also be defectors in both directions. Some of the old fogies die. Eventually one side or the other wins. If the heretics win, they become the new scientific establishment, and they defend their new paradigm just as blindly as the old establishment defended theirs; if they lose, they become a footnote in scientific history. In either case, 'normal' science then resumes.

This Kuhnian view of the scientific process seems natural to many people. It appears to explain the repeated, jarring changes that science has been forcing upon modern thought, in terms of everyday human attributes and impulses with which we are all familiar: entrenched prejudices and preconceptions, blindness to any evidence that one is mistaken, the suppression of dissent by vested interests, the desire for a quiet life, and so on. And in opposition there is the rebelliousness of youth, the quest for novelty, the joy of violating taboos and the struggle for power. Another attraction of Kuhn's ideas is that he cuts scientists down to size. No longer can they claim to be noble seekers after truth who use the rational methods of conjecture, criticism and experimental testing to solve problems and create ever better explanations of the world. Kuhn reveals that they are just rival teams playing endless games for the control of territory.

The idea of a paradigm itself is unexceptionable. We do observe and understand the world through a collection of theories, and that constitutes a paradigm. But Kuhn is mistaken in thinking that holding a paradigm blinds one to the merits of another paradigm, or prevents one from switching paradigms, or indeed prevents one from comprehending two paradigms at the same time. (For a discussion of the broader implications of this error, see Popper's *The Myth of the Framework.*) Admittedly, there is always a danger that we may underestimate or entirely miss the explanatory power of a new, fundamental theory by evaluating it from within the conceptual framework of the old theory. But it is only a danger, and given enough care and intellectual integrity, we may avoid it.

It is also true that people, scientists included, and especially those in positions of power, do tend to become attached to the prevailing way of doing things, and can be suspicious of new ideas when they are quite comfortable with the old ones. No one could claim that all scientists are uniformly and scrupulously rational in their judgement of ideas. Unjustified loyalty to paradigms is indeed a frequent cause of controversy in science, as it is elsewhere. But considered as a description or analysis of the scientific process, Kuhn's theory suffers from a fatal flaw. It explains the *succession* from one paradigm to another in sociological or psychological terms, rather than as having primarily to do with the objective merit of the rival explanations. Yet unless one understands science as a quest for explanations, the fact that it does find successive explanations, each objectively better than the last, is inexplicable.

Hence Kuhn is forced flatly to deny that there has been objective improvement in successive scientific explanations, or that such improvement is possible, even in principle:

> there is [a step] which many philosophers of science wish to take and which I refuse. They wish, that is, to compare theories as representations of nature, as statements about 'what is really out there'. Granted that neither theory of a historical pair is true, they nonetheless seek a sense in which the later is a better approximation to the truth. I believe that nothing of the sort can be found. (in Lakatos and Musgrave (eds), *Criticism and the Growth of Knowledge*, p. 265)

So the growth of objective scientific knowledge cannot be explained in the Kuhnian picture. It is no good trying to pretend that successive explanations are better only in terms of their own paradigm. There are objective differences. We can fly, whereas for most of human history people could only dream of this. The ancients would not have been blind to the efficacy of our flying machines just because, within their paradigm, they could not conceive of how they work. The reason why we can fly is that we understand 'what is really out there' well enough to build flying machines. The reason why the ancients could not is that their understanding was objectively inferior to ours.

If one does graft the reality of objective scientific progress onto Kuhn's theory, it then implies that the entire burden of fundamental innovation is carried by a handful of iconoclastic geniuses. The rest of the scientific community have their uses, but in significant matters they only hinder the growth of knowledge. This romantic view (which is often advanced independently of Kuhnian ideas) does not correspond with reality either. There have indeed been geniuses who have single-handedly revolutionized entire sciences; several have been mentioned in this book – Galileo, Newton,

Faraday, Darwin, Einstein, Gödel, Turing. But on the whole, these people managed to work, publish and gain recognition *despite* the inevitable opposition of stick-in-the-muds and time-servers. (Galileo was brought down, but not by rival scientists.) And though most of them did encounter irrational opposition, none of their careers followed the iconoclast-versus-scientific-establishment stereotype. Most of them derived benefit and support from their interactions with scientists of the previous paradigm.

I have sometimes found myself on the minority side of fundamental scientific controversies. But I have never come across anything like a Kuhnian situation. Of course, as I have said, the majority of the scientific community is not always quite as open to criticism as it ideally should be. Nevertheless, the extent to which it adheres to 'proper scientific practice' in the conduct of scientific research is nothing short of remarkable. You need only attend a research seminar in any fundamental field in the 'hard' sciences to see how strongly people's behaviour *as researchers* differs from human behaviour in general. Here we see a learned professor, acknowledged as the leading expert in the entire field, delivering a seminar. The seminar room is filled with people from every rank in the hierarchy of academic research, from graduate students who were introduced to the field only weeks ago, to other professors whose prestige rivals that of the speaker. The academic hierarchy is an intricate power structure in which people's careers, influence and reputation are continuously at stake, as much as in any cabinet room or boardroom – or more so. Yet so long as the seminar is in progress it may be quite hard for an observer to distinguish the participants' ranks. The most junior graduate student asks a question: 'Does your third equation really follow from the second one? Surely that term you omitted is not negligible.' The professor is sure that the term *is* negligible, and that the student is making an error of judgement that someone more experienced would not have made. So what happens next?

In an analogous situation, a powerful chief executive whose business judgement was being contradicted by a brash new recruit might say, 'Look, I've made more of these judgements than you've had hot dinners. If I tell you it works, then it works.' A senior politician might say in response to criticism from an obscure but ambitious party worker, 'Whose side are you on, anyway?' Even our professor, *away from the research context* (while delivering an undergraduate lecture, say) might well reply dismissively, 'You'd better learn to walk before you can run. Read the textbook, and meanwhile don't waste your time and ours.' But in the research seminar any such response to criticism would cause a wave of embarrassment to pass through the seminar room. People would avert their eyes and pretend to be diligently studying their notes. There would be smirks and sidelong glances. Everyone would be shocked by the sheer impropriety of such an attitude. In this situation, appeals to authority (at least, overt ones) are simply not acceptable, even when the most senior person in the entire field is addressing the most junior.

So the professor takes the student's point seriously, and responds with a concise but adequate argument in defence of the disputed equation. The professor tries hard to show no sign of being irritated by criticism from so lowly a source. *Most* of the questions from the floor will have the form of criticisms which, if valid, would diminish or destroy the value of the professor's life's work. But bringing vigorous and diverse criticism to bear on accepted truths is one of the very purposes of the seminar. Everyone takes it for granted that the truth is not obvious, and that the obvious need not be true; that ideas are to be accepted or rejected according to their content and not their origin; that the greatest minds can easily make mistakes; and that the most trivial-seeming objection may be the key to a great new discovery.

So the participants in the seminar, while they are engaged in science, do behave in large measure with scientific rationality. But now the seminar ends. Let us follow the group into the dining-hall. Immediately, normal human social behaviour reasserts itself. The professor is treated with deference, and sits at a table with those of equal rank. A chosen few from the lower ranks are given the privilege of being allowed to sit there too. The conversation turns to the weather, gossip or (especially) academic politics. So long as those subjects are being discussed, all the dogmatism and prejudice, the pride and loyalty, the threats and flattery of typical human interactions in similar circumstances will reappear. But if the conversation happens to revert to the subject of the seminar, the scientists instantly become scientists again. Explanations are sought, evidence and argument rule, and rank becomes irrelevant to the course of the argument. That is, at any rate, my experience in the fields in which I have worked.

Even though the history of quantum theory provides many examples of scientists clinging irrationally to what could be called 'paradigms', it would be hard to find a more spectacular counter-example to Kuhn's theory of paradigm *succession*. The discovery of quantum theory was undoubtedly a conceptual revolution, perhaps the greatest since Galileo, and there were indeed some 'old fogies' who never accepted it. But the major figures in physics, including almost all those who could be considered part of the physics establishment, were immediately ready to drop the classical paradigm. It rapidly became common ground that the new theory required a radical departure from the classical conception of the fabric of reality. The only debate was about what the new conception must be. After a while, a new orthodoxy was established by the physicist Niels Bohr and his 'Copenhagen school'. This new orthodoxy was never accepted widely enough *as a description of reality* for it to be called a paradigm, though overtly it was endorsed by most physicists (Einstein was a notable exception). Remarkably, it was not centred on the proposition that the new quantum theory was true. On the contrary, it depended crucially on quantum theory, at least in its current form, being false! According to the 'Copenhagen interpretation', the equations of quantum theory apply only to unobserved aspects of physical reality. At moments of observation a different type of process takes over, involving a direct interaction between human consciousness and subatomic physics. One particular state of consciousness becomes real, the rest were only possibilities. The Copenhagen interpretation specified this alleged process only in outline; a fuller description was deemed to be a task for the future, or perhaps to be forever beyond human comprehension. As for the unobserved events that interpolated between conscious observations, one was 'not permitted to ask' about them! How physicists, even during what was the heyday of positivism and instrumentalism, could accept such an insubstantial construction as the orthodox version of a fundamental theory is a question for historians. We need not concern ourselves here with the arcane details of the Copenhagen interpretation, because its motivation was essentially to avoid the conclusion that reality is multi-valued, and for that reason alone it is incompatible with any genuine explanation of quantum phenomena.

Some twenty years later, Hugh Everett, then a Princeton graduate student working under the eminent physicist John Archibald Wheeler, first set out the many-universes implications of quantum theory. Wheeler did not accept them. He was (and still is) convinced that Bohr's vision, though incomplete, was the basis of the correct explanation. But did he therefore behave as the Kuhnian stereotype would lead us to expect? Did he try to suppress his student's heretical ideas? On the contrary, Wheeler was afraid that Everett's ideas might not be sufficiently appreciated. So he himself wrote a short paper to accompany the one that Everett published, and they appeared on consecutive pages of the journal *Reviews of Modern Physics*. Wheeler's paper explained and defended Everett's so effectively that many readers assumed that they were jointly responsible for the content. Consequently the multiverse theory was mistakenly known as the 'Everett–Wheeler theory' for many years afterwards, much to Wheeler's chagrin.

Wheeler's exemplary adherence to scientific rationality may be extreme, but it is by no means unique. In this regard I must mention Bryce DeWitt, another eminent physicist who initially opposed Everett. In a historic exchange of letters, DeWitt put forward a series of detailed technical objections to Everett's theory, each of which Everett rebutted. DeWitt ended his argument on an informal note, pointing out that he just couldn't feel himself 'split' into multiple, distinct copies every time a decision was made. Everett's reply echoed the dispute between Galileo and the Inquisition. 'Do you feel the Earth move?' he asked – the point being that quantum theory *explains why* one does not feel such splits, just as Galileo's theory of inertia explains why one does not feel the Earth move. DeWitt conceded.

Nevertheless, Everett's discovery did not gain broad acceptance. Unfortunately, in the generation between the Copenhagen interpretation and Everett most physicists had given up on the idea of explanation in quantum theory. As I said, it was the heyday of positivism in the philosophy of science. Rejection (or incomprehension) of the Copenhagen interpretation, coupled with what might be called *pragmatic instrumentalism*, became (and remains) the typical physicist's attitude to the deepest known theory of reality. If instrumentalism is the doctrine that explanations are pointless because a theory is only an 'instrument' for making predictions, pragmatic instrumentalism is the practice of using scientific theories without knowing or caring what they mean. In this respect, Kuhnian pessimism about scientific rationality was borne out. But the Kuhnian story of how new paradigms replace old ones was not borne out at all. In a sense, pragmatic instrumentalism itself became a 'paradigm' which physicists adopted to replace the classical idea of an objective reality. But this is not the sort of paradigm that one understands the world through! In any case, whatever else physicists were doing they were not viewing the world through the paradigm of classical physics – the epitome, among other things, of objective realism and determinism. Most of them dropped it almost as soon as quantum theory was proposed, even though it had held sway over the whole of science, unchallenged ever since Galileo won the intellectual argument against the Inquisition a third of a millennium earlier.

Pragmatic instrumentalism has been feasible only because, in most branches of physics, quantum theory is not applied in its explanatory capacity. It is used only indirectly, in the testing of other theories, and only its predictions are needed. Thus generations of physicists have found it sufficient to regard interference processes, such as those that take place for a thousand-trillionth of a second when two elementary particles collide, as a 'black box': they prepare an input, and they observe an output. They use the equations of quantum theory to predict the one from the other, but they neither know nor care *how the output comes about* as a result of the input. However, there are two branches of physics where this attitude is impossible because the internal workings of the quantum-mechanical object constitute the entire subject-matter of that branch. Those branches are the quantum theory of computation, and quantum cosmology (the quantum theory of physical reality as a whole). After all, it would be a poor 'theory of computation' that never addressed issues of how the output is obtained from the input! And as for quantum cosmology, we can neither prepare an input at the beginning of the multiverse nor measure an output at the end. Its internal workings are all there is. For this reason, quantum theory is used in its full, multiverse form by the overwhelming majority of researchers in these two fields.

So Everett's story is indeed that of an innovative young researcher challenging a prevailing consensus and being largely ignored until, decades later, his view gradually becomes the new consensus. But the basis of Everett's innovation was not a claim that the prevailing theory is false, but that it is true! The incumbents, far from being able to think only in terms of their own theory, were refusing to think in its terms, and were using it only instrumentally. Yet they had dropped the previous explanatory paradigm, classical physics, with scarcely a complaint as soon as a better theory was available.

Something of the same strange phenomenon has also occurred in the other three theories that provide the main strands of explanation of the fabric of reality: the theories of computation, evolution and knowledge. In all cases the theory that now prevails, though it has definitely displaced its predecessor and other rivals in the sense that it is being applied routinely in pragmatic ways, has nevertheless failed to become the new 'paradigm'. That is, it has not been taken on board as a fundamental explanation of reality by those who work in the field.

The Turing principle, for instance, has hardly ever been seriously doubted as a pragmatic truth, at least in its weak forms (for example, that a universal computer could render any physically possible environment). Roger Penrose's criticisms are a rare exception, for he understands that contradicting the Turing principle involves contemplating radically new theories in both physics and epistemology, and some interesting new assumptions about biology too. Neither Penrose nor anyone else has yet actually proposed any viable rival to the Turing principle, so it remains the prevailing fundamental theory of computation. Yet the proposition that *artificial intelligence* is possible in principle, which follows by simple logic from this prevailing theory, is by no means taken for granted. (An artificial intelligence is a computer program that possesses properties of the human mind including intelligence, consciousness, free will and emotions, but runs on hardware other than the human brain.) The possibility of artificial intelligence is bitterly contested by eminent philosophers (including, alas, Popper), scientists and mathematicians, and by at least one prominent computer scientist. But few of these opponents seem to understand that they are contradicting the acknowledged fundamental principle of a fundamental discipline. They contemplate no alternative foundations for the discipline, as Penrose does. It is as if they were denying the possibility that we could travel to Mars, without noticing that our best theories of engineering and physics say that we can. Thus they violate a basic tenet of rationality – that good explanations are not to be discarded lightly.

But it is not only the opponents of artificial intelligence who have failed to incorporate the Turing principle into their paradigm. Very few others have done so either. The fact that four decades passed after the principle was proposed before anyone investigated its implications for physics, and a further decade passed before quantum computation was discovered, bears witness to this. People were accepting and using the principle pragmatically within computer science, but it was not integrated with their overall world-view.

Popper's epistemology has, in every pragmatic sense, become the prevailing theory of the nature and growth of scientific knowledge. When it comes to the rules for experiments in any field to be accepted as 'scientific evidence' by theoreticians in that field, or by respectable journals for publication, or by physicians for choosing between rival medical treatments, the modern watchwords are just as Popper would have them: experimental testing, exposure to criticism, theoretical explanation and the acknowledgement of fallibility in experimental procedures. In popular accounts of science, scientific theories tend to be presented more as bold conjectures than as inferences drawn from accumulated data, and the difference between science and (say) astrology is correctly explained in terms of testability rather than degree of confirmation. In school laboratories, 'hypothesis formation and testing' are the order of the day. No longer are pupils expected to 'learn by experiment', in the sense that I and my contemporaries were – that is, we were given some equipment and told what to do with it, but we were not told the theory that the results were supposed to conform to. It was hoped that we would induce it.

Despite being the prevailing theory in that sense, Popperian epistemology forms part of the world-view of very few people. The popularity of Kuhn's theory of the succession of paradigms is one illustration of this. More seriously, very few philosophers agree with Popper's claim that there is no longer a 'problem of induction' because we do not in fact obtain or justify theories from observations, but proceed by explanatory conjectures and refutations instead. It is not that many philosophers are inductivists, or have much disagreement with Popper's description and prescription of scientific method, or believe that scientific theories are actually unsound because of their conjectural status. It is that they do not accept Popper's *explanation* of how it all works. Here, again, is an echo of the Everett story. The majority view is that there is a fundamental philosophical problem with the Popperian methodology, even though science (wherever it was successful) has always followed it. Popper's heretical innovation takes the form of a claim that the methodology has been valid all along.

Darwin's theory of evolution is also the prevailing theory in its field, in the sense that no one seriously doubts that evolution through natural selection, acting on populations with random variations, is the 'origin of species' and of biological adaptation in general. No serious biologist or philosopher attributes the origin of species to divine creation or to Lamarckian evolution. (Lamarckism, an evolutionary theory that Darwinism superseded, was the analogue of inductivism. It attributed biological adaptations to the inheritance of characteristics that the organism had striven for and acquired during its life.) Yet, just as with the other three strands, objections to pure Darwinism *as an explanation* of the phenomena of the biosphere are numerous and widespread. One class of objections centres on the question whether in the history of the biosphere there has been enough time for such colossal complexity to have evolved by natural selection alone. No viable rival theory has been advanced to substantiate such objections, except conceivably the idea, of which the astronomers Fred Hoyle and Chandra Wickramasinghe are recent proponents, that the complex molecules on which life is based originated in outer space. But the point of such objections is not so much to contradict the Darwinian model as to claim that something fundamental remains unexplained in the matter of how the adaptations we observe in the biosphere came into being.

Darwinism has also been criticized as being circular because it invokes 'the survival of the fittest' as an explanation, while the 'fittest' are defined retrospectively, by their having survived. Alternatively, in terms of an independent definition of 'fitness', the idea that evolution 'favours the fittest' seems to be contradicted by the facts. For example, the most intuitive definition of biological fitness would be 'fitness of a species for survival in a particular niche', in the sense that a tiger might be thought to be the optimal machine for occupying the ecological niche that tigers occupy. The standard counter-examples to that sort of 'survival of the fittest' are adaptations, such as the peacock's tail, that seem to make the organism much *less* fit to exploit its niche. Such objections seem to undermine the ability of Darwin's theory to meet its original purpose, which was to explain how the apparent 'design' (i.e. adaptations) in living organisms could have come into being through the operation of 'blind' laws of physics on inanimate matter, without the intervention of a purposeful Designer.

Richard Dawkins' innovation, as set out in his books *The Selfish Gene* and *The Blind Watchmaker*, is yet again the claim that the prevailing theory is true after all. He argues that none of the current objections to the unadorned Darwinian model turn out, on careful inspection, to have any substance. In other words, Dawkins claims that Darwin's theory of evolution does provide a full explanation of the origin of biological adaptations. Dawkins elaborated Darwin's theory in its modern form as the theory of replicators. The replicator that is best at getting itself replicated in a given environment will eventually displace all variants of itself because, by definition, they are worse at getting themselves replicated. It is not the fittest *species* variant that survives (Darwin had not quite realized this) but the fittest *gene* variant. One consequence of this is that sometimes a gene may displace variant genes (such as genes for less cumbersome tails in peacocks) by means (such as sexual selection) that do not especially promote the good of the species or the individual. But all evolution promotes the 'good' (i.e. the replication) of the best-replicating genes – hence the term 'selfish gene'. Dawkins meets each of the objections in detail and shows that Darwin's theory, correctly interpreted, has none of the alleged flaws and does indeed explain the origin of adaptations.

It is specifically Dawkins' version of Darwinism that has become the prevailing theory of evolution in the pragmatic sense. Yet it is still by no means the prevailing *paradigm*. Many biologists and philosophers are still haunted by the feeling that there is some fundamental gap in the explanation. For example, in the same sense that Kuhn's theory of 'scientific revolutions' challenges the Popperian picture of science, there is a corresponding evolutionary theory which challenges Dawkins' picture of evolution. This is the theory of *punctuated equilibrium*, which says that evolution happens in short bursts, with long periods of unselected change in between. This theory may even be factually true. It does not actually contradict the 'selfish gene' theory, any more than Popperian epistemology is contradicted by the proposition that conceptual revolutions do not happen every day, or that scientists often resist fundamental innovation. But just as with Kuhn's theory, the way in which punctuated equilibrium and other variant evolutionary scenarios have been presented, as solving some allegedly overlooked problem in the prevailing evolutionary theory, reveals the extent to which the explanatory power of Dawkins' theory has yet to be assimilated.

There has been a very unfortunate consequence, for all four strands, of the prevailing theory's being generally rejected as an explanation, without serious rival explanations being current. It is that the proponents of the prevailing theories – Popper, Turing, Everett, Dawkins and their supporters – have found themselves constantly on the defensive against obsolete theories. The debate between Popper and most of his critics was (as I said in Chapters 3 and 7) effectively about the problem of induction. Turing spent the last years of his life in effect defending the proposition that human brains do not operate by supernatural means. Everett left scientific research after making no headway, and for several years the theory of the multiverse was championed almost single-handedly by Bryce DeWitt until progress in quantum cosmology in the 1970s forced its pragmatic acceptance in that field. But the opponents of the multiverse theory *as an explanation* have seldom advanced rival explanations. (David Bohm's theory, which I mentioned in Chapter 4, is an exception.) Instead, as the cosmologist Dennis Sciama once remarked, 'When it comes to the interpretation of quantum mechanics, the standard of argument suddenly drops to zero.' Proponents of the multiverse theory typically face a wistful, defiant but incoherent appeal to the Copenhagen interpretation – which, however, hardly anyone still believes. And finally, Dawkins has somehow become the public defender of scientific rationality against, of all things, *creationism*, and more generally against a pre-scientific world-view that has been obsolete since Galileo. The frustrating thing about all this is that, so long as the proponents of our best theories of the fabric of reality have to expend their intellectual energies in futile refutation and re-refutation of theories long known to be false, the state of our deepest knowledge cannot improve. Either Turing or Everett could easily have discovered the quantum theory of computation. Popper could have been elaborating the theory of scientific explanation. (In fairness I must acknowledge that he did understand and elaborate some connections between his epistemology and the theory of evolution.) Dawkins could, for instance, be advancing his own theory of the evolution of replicating ideas (memes).

The unified theory of the fabric of reality that is the subject of this book is, at the most straightforward level, merely the combination of the four prevailing fundamental theories of their respective fields. In that sense it too is the 'prevailing theory' of those four fields taken as a whole. Even some of the connections between the four strands are quite widely acknowledged. My thesis, therefore, also takes the form 'the prevailing theory is true after all!' Not only do I advocate taking each of the fundamental theories seriously as an explanation of its own subject-matter, I argue that taken together they provide a new level of explanation of a unified fabric of reality.

I have also argued that none of the four strands can be properly understood independently of the other three. This is possibly a clue to the reason why all these prevailing theories have not been believed. All four individual explanations share an unattractive property which has been variously criticized as 'idealized and unrealistic', 'narrow' or 'naïve' – and also 'cold', 'mechanistic' and 'lacking in humanity'. I believe that there is some truth in the gut feeling behind these criticisms. For example, of those who deny the possibility of artificial intelligence, and find themselves in effect denying that the brain is a physical object, a few are really only trying to express a much more reasonable criticism: that the Turing explanation of computation seems to leave no room, even in principle, for any future explanation *in physical terms* of mental attributes such as consciousness and free will. It is then not good enough for artificial-intelligence enthusiasts to respond brusquely that the Turing principle guarantees that a computer can do everything a brain can do. That is of course true, but it is an answer in terms of prediction, and the problem is one of explanation. There is an *explanatory gap*.

I do not believe that this gap can be filled without bringing in the other three strands. Now, as I have said, my guess is that the brain is a classical computer and not a quantum computer, so I do not expect the explanation of consciousness to be that it is any sort of quantum-computational phenomenon. Nevertheless, I expect the unification of computation and quantum physics, and probably the wider unification of all four strands, to be essential to the fundamental *philosophical* advances from which an understanding of consciousness will one day flow. Lest the reader find this paradoxical, let me draw an analogy with a similar problem from an earlier era, 'What is life?' This problem was solved by Darwin. The essence of the solution was the idea that the intricate and apparently purposeful design that is apparent in living organisms is not built into reality *ab initio*, but is an emergent consequence of the operation of the laws of physics. The laws of physics had not specifically mandated the shapes of elephants and peacocks, any more than a Creator had. They make no reference to outcomes, especially emergent ones; they merely determine the rules under which atoms and the like interact. Now, this conception of a law of nature as a set of laws of motion is relatively recent. It can, I think, be credited specifically to Galileo, and to some extent to Newton. The previous concept of a law of nature had been that of a rule stating *what happens*. An example is Johannes Kepler's laws of planetary motion, which described how the planets move in elliptical orbits. This is to be contrasted with Newton's laws, which are laws of physics in the modern sense. They make no mention of ellipses, though they reproduce (and correct) Kepler's predictions under appropriate conditions. No one could have explained what life is under Kepler's conception of a 'law of physics', for they would have been looking for a law that mandates elephants in the same way as Kepler's laws mandate ellipses. But Darwin was able to wonder how laws of nature that did not mention elephants could nevertheless produce them, just as Newton's laws produce ellipses. Although Darwin made no use of any specific law of Newton's, his discovery would have been inconceivable without the world-view underlying those laws. That is the sense in which I expect the solution of the 'What is consciousness?' problem to depend on quantum theory. It will invoke no specific quantum-mechanical processes, but it will depend crucially on the quantum-mechanical, and especially the multi-universe, world-picture.

What is my evidence? I have already presented some of it in Chapter 8, where I discussed the multiverse view of knowledge. Although we do not know what consciousness is, it is clearly intimately related to the growth and representation of knowledge within the brain. It seems unlikely, then, that we shall be able to explain what consciousness is, as a physical process, before we have explained knowledge in physical terms. Such an explanation has been elusive in the classical theory of computation. But, as I explained, in quantum theory there is a good basis for one: knowledge can be understood as complexity that extends across large numbers of universes.

Another mental attribute that is somehow associated with consciousness is free will. Free will is also notoriously difficult to understand in the classical world-picture. The difficulty of reconciling free will with physics is often attributed to determinism, but it is not determinism that is at fault. It is (as I have explained in Chapter 11) classical spacetime. In spacetime, *something* happens to me at each particular moment in my future. Even if what will happen is unpredictable, it is already there, on the appropriate cross-section of spacetime. It makes no sense to speak of my 'changing' what is on that cross-section. Spacetime does not change, therefore one cannot, within spacetime physics, conceive of causes, effects, the openness of the future or free will.

Thus, replacing deterministic laws of motion by indeterministic (random) ones would do nothing to solve the problem of free will, so long as the laws remained classical. Freedom has nothing to do with randomness. We value our free will as the ability to express, in our actions, who we as individuals are. Who would value being random? What we think of as our *free* actions are not those that are random or undetermined but those that are largely *determined* by who we are, and what we think, and what is at issue. (Although they are largely determined, they may be highly unpredictable in practice for reasons of complexity.)

Consider this typical statement referring to free will: 'After careful thought I chose to do X; I could have chosen otherwise; it was the right decision; I am good at making such decisions.' In any classical world-picture this statement is pure gibberish. In the multiverse picture it has a straightforward physical representation, shown in Table 13.1. (I am not proposing to *define* moral or aesthetic values in terms of such representations; I am merely pointing out that, thanks to the multiverse character of quantum reality, free will and related concepts are now compatible with physics.)

Thus Turing's conception of computation seems less disconnected from human values, and is no obstacle to the understanding of human attributes like free will, provided it is understood in a multiverse context. The same example exonerates Everett's theory itself. On the face of it, the price of understanding interference phenomena is to create or exacerbate many philosophical problems. But here, and in many other examples I have given in this book, we see that the very opposite is the case. The fruitfulness of the multiverse theory in contributing to the solution of long-standing philosophical problems is so great that it would be worth adopting even if there were no physical evidence for it at all. Indeed, the

After careful thought I chose to do X	After careful thought some copies of me, including the one speaking, chose to do X
I could have chosen otherwise	Other copies of me chose otherwise
It was the right decision	Representations of the moral or aesthetic values that are reflected in my choice of option X are repeated much more widely in the multiverse than representations of rival values
I am good at making such decisions	The copies of me who chose X, and who chose rightly in other such situations, greatly outnumber those who did not

TABLE 13.1 *Physical representations of some statements referring to free will.*

philosopher David Lewis, in his book *On the Plurality of Worlds*, has postulated the existence of a multiverse for philosophical reasons alone.

Turning again to the theory of evolution, I can similarly attribute *some* sense to those who criticize Darwinian evolution on the grounds that it seems 'unlikely' that such complex adaptations could have evolved in the given time. One of Dawkins' critics wants us to be as surprised by the biosphere as we would be if a heap of spare parts thrown together happened to fall into the pattern of a Boeing 747. On the face of it, this critic is forcing an analogy between, on the one hand, billions of years of planet-wide trial and error, and on the other hand an instantaneous accident of 'happening to fall together'. That would be wilfully to miss the whole point of the evolutionary explanation. Nevertheless, is Dawkins' precisely opposite position completely adequate as an explanation? Dawkins wants us *not to be surprised* that complex adaptations have come into being spontaneously. In other words, he is claiming that his 'selfish gene' theory is a full explanation – not of course for specific adaptations, but of how it was possible for such complex adaptations to come into being.

But it is not a full explanation. There is an explanatory gap, and this time we already know much more about how the other strands could fill it. We have seen that the very fact that physical variables can store information, that they can interact with one another to transfer and replicate it, and that such processes are stable, all depend on the details of quantum theory. Furthermore, we have seen that the existence of highly adapted replicators depends on the physical feasibility of virtual-reality generation and universality, which in turn can be understood as consequences of a deep principle, the Turing principle, that links physics and the theory of computation and makes no explicit reference to replicators, evolution or biology at all.

An analogous gap exists in Popperian epistemology. Its critics wonder why the scientific method works, or what justifies our reliance on the best scientific theories. This leads them to hanker after a principle of induction or something of the sort (though, as crypto-inductivists, they usually realize that such a principle would not explain or justify anything either). For Popperians to reply that there is no such thing as justification, or that it is never rational to rely on theories, is to provide no explanation. Popper even said that 'no theory of knowledge should attempt to explain why we are successful in our attempts to explain things' (*Objective Knowledge* p. 23). But, once we understand that the growth of human knowledge is a physical process, we see that it cannot be illegitimate to try to explain how and why it occurs. Epistemology is a theory of (emergent) physics. It is a factual theory about the circumstances under which a certain physical quantity (knowledge) will or will not grow. The bare assertions of this theory are largely accepted. But we cannot possibly find an explanation of why they are true solely within the theory of knowledge *per se*. In that narrow sense, Popper was right. The explanation must involve quantum physics, the Turing principle and, as Popper himself stressed, the theory of evolution.

The proponents of the prevailing theory, in each of the four cases, are put permanently on the defensive by their critics' harping on these explanatory gaps. This often forces them to retreat into the core of their own strand. 'Here I stand, I can do no other' is their ultimate response, as they rely on the self-evident irrationality of abandoning the unrivalled fundamental theory of their own particular field. This only makes them seem even more narrow to the critics, and it tends to engender pessimism about the very prospect of further fundamental explanation.

Despite all the excuses I have been making for the critics of the central theories, the history of all four strands shows that something very unpleasant happened to fundamental science and philosophy for most of the twentieth century. The popularity of positivism and of an instrumentalist view of science was connected with an apathy, loss of self-confidence and pessimism about genuine explanations at a time when the prestige, usefulness and, indeed, funding for fundamental research were all at an all-time high. Of course there were many individual exceptions, including the four heroes of this chapter. But the unprecedented manner in which their theories were simultaneously adopted and ignored speaks for itself. I do not claim to have a full explanation for this phenomenon, but whatever caused it, we seem to be coming out of it now.

I have pointed out one possible contributory cause, namely that individually, all four theories have explanatory gaps that can make them seem narrow, inhuman and pessimistic. But I suggest that when they are taken together as a unified explanation of the fabric of reality, this unfortunate property is reversed. Far from denying free will, far from placing human values in a context where they are trivial and insignificant, far from being pessimistic, it is a fundamentally optimistic world-view that places human minds at the centre of the physical universe, and explanation and understanding at the centre of human purposes. I hope we shall not have to spend too long looking backwards to defend this unified view against non-existent competitors. There will be no lack of competitors when, having taken the unified theory of the fabric of reality seriously, we begin to develop it further. It is time to move on.

TERMINOLOGY

paradigm The set of ideas through which those who hold it observe and explain everything in their experience.
According to Thomas Kuhn, holding a paradigm blinds one to the merits of another paradigm and prevents one from switching paradigms. One cannot comprehend two paradigms at the same time.

Copenhagen interpretation of quantum mechanics An idea for making it easier to evade the implications of quantum theory for the nature of reality. At moments of observation, the outcome in one of the universes supposedly becomes real, and all the other universes – even those that contributed to that outcome – are deemed never to have existed. Under this view, one is not permitted to ask about what happens in reality between conscious observations.

SUMMARY

The intellectual histories of the fundamental theories of the four strands contain remarkable parallels. All four have been simultaneously accepted (for use in practice) and ignored (as explanations of reality). One reason for this is that, taken individually, each of the four theories has explanatory gaps, and seems cold and pessimistic. To base a world-view on any of them individually is, in a generalized sense, reductionist. But when they are taken together as a unified explanation of the fabric of reality, this is no longer so.

Whatever next?

14

The Ends of the Universe

Although history has no meaning, we can give it a meaning.
Karl Popper (*The Open Society and Its Enemies*, Vol. 2, p. 278)

When, in the course of my research on the foundations of quantum theory, I was first becoming aware of the links between quantum physics, computation and epistemology, I regarded these links as evidence of the historical tendency for physics to swallow up subjects that had previously seemed unrelated to it. Astronomy, for example, was linked with terrestrial physics by Newton's laws, and over the next few centuries much of it was absorbed and became astrophysics. Chemistry began to be subsumed into physics by Faraday's discoveries in electrochemistry, and quantum theory has made a remarkable proportion of basic chemistry directly predictable from the laws of physics alone. Einstein's general relativity swallowed geometry, and rescued both cosmology and the theory of time from their former purely philosophical status, making them into fully integrated branches of physics. Recently, as I have discussed, the theory of time travel has been integrated as well.

Thus, the further prospect of quantum physics absorbing not only the theory of computation but also, of all things, *proof theory* (which has the alternative name 'meta-mathematics') seemed to me to be evidence of two trends. First, that human knowledge as a whole was continuing to take on the unified structure that it would have to have if it was comprehensible in the strong sense I hoped for. And second, that the unified structure itself was going to consist of an ever deepening and broadening theory of fundamental physics.

The reader will know that I have changed my mind about the second point. The character of the fabric of reality that I am now proposing is not that of fundamental physics alone. For example, the quantum theory of computation has not been constructed by deriving principles of computation from quantum physics alone. It includes the Turing principle, which was already, under the name of the Church–Turing *conjecture*, the basis of the theory of computation. It had never been used in physics, but I have argued that it is only as a principle of physics that it can be properly understood. It is on a par with the principle of the conservation of energy and the other laws of thermodynamics: that is, it is a constraint that, to the best of our knowledge, all other theories conform to. But, unlike existing laws of physics, it has an emergent character, referring directly to the properties of complex machines and only consequentially to subatomic objects and processes. (Arguably, the second law of thermodynamics – the principle of increasing entropy – is also of that form.)

Similarly, if we understand *knowledge* and *adaptation* as structure which extends across large numbers of universes, then we expect the principles of epistemology and evolution to be expressible directly as laws about the structure of the multiverse. That is, they are physical laws, but at an emergent level. Admittedly, quantum complexity theory has not yet reached the point where it can express, in physical terms, the proposition that knowledge can grow only in situations that conform to the Popperian pattern shown in Figure 3.3. But that is just the sort of proposition that I expect to appear in the nascent Theory of Everything, the unified explanatory and predictive theory of all four strands.

That being so, the view that quantum physics is swallowing the other strands must be regarded merely as a narrow, physicist's perspective, tainted, perhaps, by reductionism. Indeed, each of the other three strands is quite rich enough to form the whole foundation of some people's world-view in much the same way that fundamental physics forms the foundation of a reductionist's world-view.

Richard Dawkins thinks that 'If superior creatures from space ever visit Earth, the first question they will ask, in order to assess the level of our civilisation, is: "Have they discovered evolution yet?"' Many philosophers have agreed with René Descartes that epistemology underlies all other knowledge, and that something like Descartes's *cogito ergo sum* argument is our most basic explanation. Many computer scientists have been so impressed with recently discovered connections between physics and computation that they have concluded that the universe *is* a computer, and the laws of physics are programs that run on it. But all these are narrow, even misleading perspectives on the true fabric of reality. Objectively, the new synthesis has a character of its own, substantially different from that of any of the four strands it unifies.

For example, I have remarked that the fundamental theories of each of the four strands have been criticized, in part justifiably, for being 'naïve', 'narrow', 'cold', and so on. Thus, from the point of view of a reductionist physicist such as Stephen Hawking, the human race is just an astrophysically insignificant 'chemical scum'. Steven Weinberg thinks that 'The more the universe seems comprehensible, the more it also seems pointless. But if there is no solace in the fruits of our research, there is at least some consolation in the research itself.' (*The First Three Minutes*, p. 154.) But anyone not involved in fundamental physics must wonder why.

As for computation, the computer scientist Tomasso Toffoli has remarked that 'We never perform a computation ourselves, we just hitch a ride on the great Computation that is going on already.' To him, this is no cry of despair – quite the contrary. But critics of the computer-science world-view do not want to see themselves as just someone else's program running on someone else's computer. Narrowly conceived evolutionary theory considers us mere 'vehicles' for the replication of our genes or memes; and it refuses to address the question of why evolution has tended to create ever greater adaptive complexity, or the role that such complexity plays in the wider scheme of things. Similarly, the (crypto-)inductivist critique of Popperian epistemology is that, while it states the conditions for scientific knowledge to grow, it seems not to explain *why* it grows – why it creates theories that are worth using.

As I have explained, the defence in each case depends on adducing explanations from some of the other strands. We are not *merely* 'chemical scum', because (for instance) the gross behaviour of our planet, star and galaxy depend on an emergent but fundamental physical quantity: the *knowledge* in that scum. The creation of useful knowledge by science, and adaptations by evolution, must be understood as the emergence of the self-similarity that is mandated by a principle of physics, the Turing principle. And so on.

Thus the problem with taking any of these fundamental theories individually as the basis of a world-view is that they are each, in an extended sense, reductionist. That is, they have a monolithic explanatory structure in which everything follows from a few extremely deep ideas. But that leaves aspects of the subject entirely unexplained. In contrast, the explanatory structure that they *jointly* provide for the fabric of reality is not hierarchical: each of the four strands contains principles which are 'emergent' from the perspective of the other three, but nevertheless help to explain them.

Three of the four strands seem to rule out human beings and human values from the fundamental level of explanation. The fourth, epistemology, makes knowledge primary but gives no reason to regard epistemology itself as having relevance beyond the psychology of our own species. Knowledge seems a parochial concept until we consider it from a multiverse perspective. But if knowledge is of fundamental significance, we may ask what sort of role now seems natural for knowledge-creating beings such as ourselves in the unified fabric of reality. This question has been explored by the cosmologist Frank Tipler. His answer, the *omega-point theory*, is an excellent example of a theory which is, in the sense of this book, about the fabric of reality as a whole. It is not framed within any one strand, but belongs irreducibly to all four. Unfortunately Tipler himself, in his book *The Physics of Immortality*, makes exaggerated claims for his theory which have caused most scientists and philosophers to reject it out of hand, thereby missing the valuable core idea which I shall now explain.

From my own perspective, the simplest point of entry to the omega-point theory is the Turing principle. A universal virtual-reality generator is physically possible. Such a machine is able to render any physically possible environment, as well as certain hypothetical and abstract entities, to any desired accuracy. Its computer therefore has a potentially unlimited requirement for additional memory, and may run for an unlimited number of steps. This was trivial to arrange in the classical theory of computation, so long as the universal computer was thought to be purely abstract. Turing simply postulated an infinitely long memory tape (with, as he thought, self-evident properties), a perfectly accurate processor requiring neither power nor maintenance, and unlimited time available. Making the model more realistic by allowing for periodic maintenance raises no problem of principle, but the other three requirements – unlimited memory capacity, and an unlimited running time and energy supply – are problematic in the light of existing cosmological theory. In some current cosmological models, the universe will recollapse in a Big Crunch after a finite time, and is also spatially finite. It has the geometry of a '3-sphere', the three-dimensional analogue of the two-dimensional surface of a sphere. On the face of it, such a cosmology would place a finite bound on both the memory capacity and the number of processing steps the machine could perform before the universe ended. This would make a universal computer physically impossible, so the Turing principle would be violated. In other cosmological models the universe continues to expand for ever and is spatially infinite, which might seem to allow for an unlimited source of material for the manufacture of additional memory. Unfortunately, in most such models the density of energy available to power the computer would diminish as the universe expanded, and would have to be collected from ever further afield. Because physics imposes an absolute speed limit, the speed of light, the computer's memory accesses would have to slow down and the net effect would again be that only a finite number of computational steps could be performed.

The key discovery in the omega-point theory is that of a class of cosmological models in which, though the universe is finite in both space and time, the memory capacity, the number of possible computational steps and the effective energy supply are all unlimited. This apparent impossibility can happen because of the extreme violence of the final moments of the universe's Big Crunch collapse. Spacetime singularities, like the Big Bang and the Big Crunch, are seldom tranquil places, but this one is far worse than most. The shape of the universe would change from a 3-sphere to the three-dimensional analogue of the surface of an ellipsoid. The degree of deformation would increase, and then decrease, and then increase again more rapidly with respect to a different axis. Both the amplitude and frequency of these oscillations would increase without limit as the final singularity was approached, so that a literally infinite number of oscillations would occur even though the end would come within a finite time. Matter as we know it would not survive: all matter, and even the atoms themselves, would be wrenched apart by the gravitational shearing forces generated by the deformed spacetime. However, these shearing forces would also provide an unlimited source of available energy, which could in principle be used to power a computer. How could a computer exist under such conditions? The only 'stuff' left to build computers with would be elementary particles and gravity itself, presumably in some highly exotic quantum states whose existence we, still lacking an adequate theory of quantum gravity, are currently unable to confirm or deny. (Observing them experimentally is of course out of the question.) If suitable states of particles and the gravitational field exist, then they would also provide an unlimited memory capacity, and the universe would be shrinking so fast that an infinite number of memory accesses would be feasible in a finite time before the end. The end-point of the gravitational collapse, the Big Crunch of this cosmology, is what Tipler calls the omega point.

Now, the Turing principle implies that there is no upper bound on the number of computational steps that are physically possible. So, given that an omega-point cosmology is (under plausible assumptions) the only type in which an infinite number of computational steps could occur, we can infer that our actual spacetime must have the omega-point form. Since all computation would cease as soon as there were no more variables capable of carrying information, we can infer that the necessary physical variables (perhaps quantum-gravitational ones) do exist right up to the omega point.

A sceptic might argue that this sort of reasoning involves a massive, unjustified extrapolation. We have experience of 'universal' computers only in a most favourable environment which does not remotely resemble the final stages of the universe. And we have experience of them performing only a finite number of computational steps, using only a finite amount of memory. How can it be valid to extrapolate from those finite numbers to infinity? In other words, how can we know that the Turing principle in its strong form is strictly true? What evidence is there that reality supports more than *approximate* universality?

This sceptic is, of course, an inductivist. Furthermore, this is exactly the type of thinking that (as I argued in the previous chapter) prevents us from understanding our best theories and improving upon them. What is or is not an 'extrapolation' depends on which *theory* one starts with. If one starts with some vague but parochial concept of what is 'normal' about the possibilities of computation, a concept uninformed by the best available explanations in that subject, then one will regard *any* application of the theory outside familiar circumstances as 'unjustified extrapolation'. But if one starts with explanations from the best available fundamental theory, then one will consider the very idea that some nebulous 'normalcy' holds in extreme situations to be an unjustified extrapolation. To understand our best theories, we must take them seriously as explanations of reality, and not regard them as mere summaries of existing observations. The Turing principle is our best theory of the foundations of computation. Of course we know only a finite number of instances confirming it – but that is true of every theory in science. There remains, and will always remain, the logical possibility that universality holds only approximately. But there is no rival theory of computation claiming that. And with good reason, for a 'principle of approximate universality' would have no explanatory power. If, for instance, we want to understand why the world *seems* comprehensible, the explanation might be that the world *is* comprehensible. Such an explanation can, and in fact does, fit in with other explanations in other fields. But the theory that the world is *half*-comprehensible explains nothing and could not possibly fit in with explanations in other fields unless *they* explained *it*. It simply restates the problem and introduces an unexplained constant, one-half. In short, what justifies assuming that the full Turing principle holds at the end of the universe, is that any other assumption spoils good explanations of what is happening here and now.

Now, it turns out that the type of oscillations of space that would make an omega point happen are highly unstable (in the manner of classical chaos) as well as violent. And they become increasingly more so, without limit, as the omega point is approached. A small deviation from the correct shape would be magnified rapidly enough for the conditions for continuing computation to be violated, so the Big Crunch would happen after only a finite number of computational steps. Therefore, to satisfy the Turing principle and attain an omega point, the universe would have to be continually 'steered' back onto the right trajectories. Tipler has shown in principle how this could be done, by manipulating the gravitational field over the whole of space. Presumably (again we would need a quantum theory of gravity to know for sure), the technology used for the stabilizing mechanisms, and for storing information, would have to be continually improved – indeed, improved an infinite number of times – as the density and stresses became ever higher without limit. This would require the continual creation of new knowledge, which, Popperian epistemology tells us, requires the presence of rational criticism and thus of intelligent entities. We have therefore inferred, just from the Turing principle and some other independently justifiable assumptions, that intelligence will survive, and knowledge will continue to be created, until the end of the universe.

The stabilization procedures, and the accompanying knowledge-creation processes, will all have to be increasingly rapid until, in the final frenzy, an infinite amount of both occur in a finite time. We know of no reason why the physical resources should not be available to do this, but one might wonder why the inhabitants should bother to go to so much trouble. Why should they continue so carefully to steer the gravitational oscillations during, say, the last second of the universe? If you have only one second left to live, why not just sit back and take it easy at last? But of course, that is a misrepresentation of the situation. It could hardly be a bigger misrepresentation. For these people's minds will be running as computer programs in computers whose physical speed is increasing without limit. Their thoughts will, like ours, be virtual-reality renderings performed by these computers. It is true that at the end of that final second the whole sophisticated mechanism will be destroyed. But we know that the subjective duration of a virtual-reality experience is determined not by the elapsed time, but by the computations that are performed in that time. In an infinite number of computational steps there is time for an infinite number of thoughts – plenty of time for the thinkers to place themselves into any virtual-reality environment they like, and to experience it for however long they like. If they tire of it, they can switch to any other environment, or to any number of other environments they care to design. Subjectively, they will not be at the final stages of their lives but at the very beginning. They will be in no hurry, for subjectively they will live for ever. With one second, or one microsecond, to go, they will still have 'all the time in the world' to do more, experience more, create more – infinitely more – than anyone in the multiverse will ever have done before then. So there is every incentive for them to devote their attention to managing their resources. In doing so they are merely preparing for their own future, an open, infinite future of which they will be in full control and on which, at any particular time, they will be only just embarking.

We may hope that the intelligence at the omega point will consist of our descendants. That is to say, of our *intellectual* descendants, since our present physical forms could not survive near the omega point. At some stage human beings would have to transfer the computer programs that are their minds into more robust hardware. Indeed, this will eventually have to be done an infinite number of times.

The mechanics of 'steering' the universe to the omega point require actions to be taken throughout space. It follows that intelligence will have to spread all over the universe in time to make the first necessary adjustments. This is one of a series of deadlines that Tipler has shown we should have to meet – and he has shown that meeting each of them is, to the best of our present knowledge, physically possible. The first deadline is (as I remarked in Chapter 8) about five billion years from now when the Sun will, if left to its own devices, become a red giant star and wipe us out. We must learn to control or abandon the Sun before then. Then we must colonize our Galaxy, then the local cluster of galaxies, and then the whole universe. We must do each of these things soon enough to meet the corresponding deadline but we must not advance so quickly that we use up all the necessary resources before we have developed the next level of technology.

I say 'we must' do all this, but that is only on the assumption that it is we who are the ancestors of the intelligence that will exist at the omega point. We need not play this role if we do not want to. If we choose not to, and the Turing principle is true, then we can be sure that someone else (presumably some extraterrestrial intelligence) will.

Meanwhile, in parallel universes, our counterparts are making the same choices. Will they all succeed? Or, to put that another way, will someone *necessarily* succeed in creating an omega point in our universe? This depends on the fine detail of the Turing principle. It says that a universal computer is physically possible, and 'possible' usually means 'actual in this or some other universe'. Does the principle require a universal computer to be built in all universes, or only in some – or perhaps in 'most'? We do not yet understand the principle well enough to decide. Some principles of physics, such as the principle of the conservation of energy, hold only over a group of universes and may under some circumstances be violated in individual universes. Others, such as the principle of the conservation of charge, hold strictly in every universe. The two simplest forms of the Turing principle would be:

(1) there is a universal computer in *all* universes; or

(2) there is a universal computer in *at least some* universes.

The 'all universes' version seems too strong to express the intuitive idea that such a computer is physically *possible*. But 'at least some universes' seems too weak since, on the face of it, if universality holds only in very few universes then it loses its explanatory power. But a 'most universes' version would require the principle to specify a particular percentage, say 85 per cent, which seems very implausible. (There are no 'natural' constants in physics, goes the maxim, except zero, one and infinity.) Therefore Tipler in effect opts for 'all universes', and I agree that this is the most natural choice, given what little we know.

That is all that the omega-point theory – or, rather, the scientific component I am defending – has to say. One can reach the same conclusion from several different starting-points in three of the four strands. One of them is the epistemological principle that *reality is comprehensible*. That principle too is independently justifiable in so far as it underlies Popperian epistemology. But its existing formulations are all too vague for categorical conclusions about, say, the unboundedness of physical representations of knowledge, to be drawn from it. That is why I prefer not to postulate it directly, but to infer it from the Turing principle. (This is another example of the greater explanatory power that is available when one considers the four strands as being jointly fundamental.) Tipler himself relies either on the postulate that life will continue for ever, or on the postulate that information processing will continue for ever. From our present perspective, neither of these postulates seems fundamental. The advantage of the Turing principle is that it is already, for reasons quite independent of cosmology, regarded as a fundamental principle of nature – admittedly not always in this strong form, but I have argued that the strong form is necessary if the principle is to be integrated into physics.

Tipler makes the point that the science of cosmology has tended to study the *past* (indeed, mainly the distant past) of spacetime. But most of spacetime lies to the future of the present epoch. Existing cosmology does address the issue of whether the universe will or will not recollapse, but apart from that there has been very little theoretical investigation of the greater part of spacetime. In particular, the lead-up to the Big Crunch has received far less study than the aftermath of the Big Bang. Tipler sees the omega-point theory as filling that gap. I believe that the omega-point theory deserves to become the prevailing theory of the future of spacetime until and unless it is experimentally (or otherwise) refuted. (Experimental refutation is possible because the existence of an omega point in our future places certain constraints on the condition of the universe today.)

Having established the omega-point scenario, Tipler makes some additional assumptions – some plausible, others less so – which enable him to fill in more details of future history. It is Tipler's quasi-religious interpretation of that future history, and his failure to distinguish that interpretation from the underlying scientific theory, that have prevented the latter from being taken seriously. Tipler notes that an infinite amount of knowledge will have been created by the time of the omega point. He then assumes that the intelligences existing in this far future will, like us, want (or perhaps need) to discover knowledge other than what is immediately necessary for their survival. Indeed, they have the potential to discover all knowledge that is physically knowable, and Tipler assumes that they will do so.

So in a sense, the omega point will be *omniscient*.

But only in a sense. In attributing properties such as omniscience or even physical existence to the omega point, Tipler makes use of a handy linguistic device that is quite common in mathematical physics, but can be misleading if taken too literally. The device is to identify a limiting point of a sequence with the sequence itself. Thus, when he says that the omega point 'knows' X, he means that X is known by some finite entity before the time of the omega point, and is never subsequently forgotten. What he does *not* mean is that there is a knowing entity literally at the end-point of gravitational collapse, for there is no physical entity there at all. Thus in the most literal sense the omega point knows nothing, and can be said to 'exist' only because some of our explanations of the fabric of reality refer to the limiting properties of physical events in the distant future.

Tipler uses the theological term 'omniscient' for a reason which will shortly become apparent; but let me note at once that in this usage it does not carry its full traditional connotation. The omega point will not know *everything*. The overwhelming majority of abstract truths, such as truths about Cantgotu environments and the like, will be as inaccessible to it as they are to us.

Now, since the whole of space will be filled with the intelligent computer, it will be *omnipresent* (though only after a certain date). Since it will be continually rebuilding itself, and steering the gravitational collapse, it can be said to be in control of everything that happens in the material universe (or multiverse, if the omega-point phenomenon happens in all universes). So, Tipler says, it will be *omnipotent*. But again, this omnipotence is not absolute. On the contrary, it is strictly limited to the available matter and energy, and is subject to the laws of physics.

Since the intelligences in the computer will be creative thinkers, they must be classified as 'people'. Any other classification, Tipler rightly argues, would be racist. And so he claims that at the omega-point limit there is an omniscient, omnipotent, omnipresent society of people. This society, Tipler identifies as God.

I have mentioned several respects in which Tipler's 'God' differs from the God or gods that most religious people believe in. There are further differences, too. For instance, the people near the omega point could not, even if they wanted to, speak to us or communicate their wishes to us, or work miracles (today). They did not create the universe, and they did not invent the laws of physics – nor could they violate those laws if they wanted to. They may listen to prayers from the present day (perhaps by detecting very faint signals), but they cannot answer them. They are (and this we can infer from Popperian epistemology) opposed to religious faith, and have no wish to be worshipped. And so on. But Tipler ploughs on, and argues that most of the core features of the God of the Judaeo-Christian religions are also properties of the omega point. Most religious people will, I think, disagree with Tipler about what the core features of their religions are.

In particular, Tipler points out that a sufficiently advanced technology will be able to resurrect the dead. It could do this in several different ways, of which the following is perhaps the simplest. Once one has enough computer power (and remember that eventually any desired amount will be available), one can run a virtual-reality rendering of the entire universe – indeed, the entire multiverse – starting at the Big Bang, with any desired degree of accuracy. If one does not know the initial state accurately enough, one can try an arbitrarily fine sampling of all possible initial states, and render them all simultaneously. The rendering may have to pause, for reasons of complexity, if the epoch being rendered gets too close to the actual time at which the rendering is being performed. But it will soon be able to continue as more computer power comes on line. To the omega-point computers, nothing is intractable. There is only 'computable' and 'non-computable', and rendering real physical environments definitely comes into the 'computable' category. In the course of this rendering, the planet Earth and many variants of it will appear. Life, and eventually human beings, will evolve. All the human beings who have ever lived anywhere in the multiverse (that is, all those whose existence was physically possible) will appear somewhere in this vast rendering. So will every extraterrestrial and artificial intelligence that could ever have existed. The controlling program can look out for these intelligent beings and, if it wants to, place them in a better virtual environment – one, perhaps, in which they will not die again, and will have all their wishes granted (or at least, all wishes that a given, unimaginably high, level of computing resources can meet). Why would it do that? One reason might be a moral one: by the standards of the distant future, the environment we live in today is extremely harsh and we suffer atrociously. It may be considered unethical not to rescue such people and give them a chance of a better life. But it would be counter-productive to place them immediately in contact with the contemporary culture at the time of resurrection: they would be instantly confused, humiliated and overwhelmed. Therefore, Tipler says, we can expect to be resurrected in an environment of a type that is essentially familiar to us, except that every unpleasant element will have been removed, and many extremely pleasant elements will have been added. In other words, heaven.

Tipler goes on in this manner to reconstitute many other aspects of the traditional religious landscape by redefining them as physical entities or processes that can plausibly be expected to exist near the omega point. Now, let us set aside the question whether the reconstituted versions are true to their religious analogues. The whole story about what these far-future intelligences will or will not do is based on a string of assumptions. Even if we concede that these assumptions are individually plausible, the overall conclusions cannot really claim to be more than informed speculation. Such speculations are worth making, but it is important to distinguish them from the argument for the existence of the omega point itself, and from the theory of the omega point's physical and epistemological properties. For *those* arguments assume no more than that the fabric of reality does indeed conform to our best theories, an assumption that can be independently justified.

As a warning against the unreliability of even informed speculation, let me revisit the ancient master builder of Chapter 1, with his pre-scientific knowledge of architecture and engineering. We are separated from him by so large a cultural gap that it would be extremely difficult for him to conceive a workable picture of our civilization. But we and he are almost contemporaries in comparison with the tremendous gap between us and the earliest possible moment of Tiplerian resurrection. Now, suppose that the master builder is speculating about the distant future of the building industry, and that by some extraordinary fluke he happens upon a perfectly accurate assessment of the technology of the present day. Then he will know, among other things, that we are capable of building structures far vaster and more impressive than the greatest cathedrals of his day. We could build a cathedral a mile high if we chose to. And we could do it using a far smaller proportion of our wealth, and less time and human effort, than he would have needed to build even a modest cathedral. So he would have been confident in predicting that by the year 2000 there would be mile-high cathedrals. He would be mistaken, and badly so, for though we have the technology to build such structures, we have chosen not to. Indeed, it now seems unlikely that such a cathedral will ever be built. Even though we supposed our near-contemporary to be right about our technology, he would have been quite wrong about our preferences. He would have been wrong because some of his most unquestioned assumptions about human motivations have become obsolete after only a few centuries.

Similarly, it may seem natural to us that the omega-point intelligences, for reasons of historical or archaeological research, or compassion, or moral duty, or mere whimsy, will eventually create virtual-reality renderings of us, and that when their experiment is over they will grant us the piffling computational resources we would require to live for ever in 'heaven'. (I myself would prefer to be allowed gradually to join their culture.) But we cannot know what they will want. Indeed, no attempt to prophesy future large-scale developments in human (or superhuman) affairs can produce reliable results. As Popper has pointed out, the future course of human affairs depends on the future growth of knowledge. And we cannot predict what specific knowledge will be created in the future – because if we could, we should by definition already possess that knowledge in the present.

It is not only scientific knowledge that informs people's preferences and determines how they choose to behave. There are also, for instance, moral criteria, which assign attributes such as 'right' and 'wrong' to possible actions. Such values have been notoriously difficult to accommodate in the scientific world-view. They seem to form a closed explanatory structure of their own, disconnected from that of the physical world. As David Hume pointed out, it is impossible logically to derive an 'ought' from an 'is'. Yet we use such values both to explain and to determine our physical actions.

The poor relation of morality is *usefulness*. Since it seems much easier to understand what is objectively useful or useless than what is objectively right or wrong, there have been many attempts to define morality in terms of various forms of usefulness. There is, for example, evolutionary morality, which notes that many forms of behaviour which we explain in moral terms, such as not committing murder, or not cheating when we cooperate with other people, have analogues in the behaviour of animals. And there is a branch of evolutionary theory, *sociobiology*, that has had some success in explaining animal behaviour. Many people have been tempted to conclude that moral explanations for human choices are just window-dressing; that morality has no objective basis at all, and that 'right' and 'wrong' are simply tags we apply to our inborn urges to behave in one way rather than another. Another version of the same explanation replaces genes by memes, and claims that moral terminology is just window-dressing for social conditioning. However, none of these explanations fits the facts. On the one hand, we do *not* tend to explain inborn behaviour – say, epileptic fits – in terms of moral choices; we have a notion of voluntary and involuntary actions, and only the voluntary ones have moral explanations. On the other hand, it is hard to think of a single inborn human behaviour – avoiding pain, engaging in sex, eating or whatever – that human beings have not under various circumstances chosen to override for moral reasons. The same is true, even more commonly, of socially conditioned behaviour. Indeed, overriding both inborn and socially conditioned behaviours is itself a characteristic human behaviour. So is explaining such rebellions in moral terms. None of these behaviours has any analogue among animals; in none of these cases can moral explanations be reinterpreted in genetic or memetic terms. This is a fatal flaw of this entire class of theories. Could there be a gene for overriding genes when one feels like it? Social conditioning that promotes rebellion? Perhaps, but that still leaves the problem of *how we choose what to do instead*, and of what we mean when we explain our rebellion by claiming that we were simply right, and that the behaviour prescribed by our genes or by our society in this situation was simply evil.

These genetic theories can be seen as a special case of a wider stratagem, that of denying that moral judgements are meaningful on the grounds that we do not really choose our actions – that free will is an illusion incompatible with physics. But in fact, as we saw in Chapter 13, free will *is* compatible with physics, and fits quite naturally into the fabric of reality that I have described.

Utilitarianism was an earlier attempt to integrate moral explanations with the scientific world-view through 'usefulness'. Here 'usefulness' was identified with human happiness. Making moral choices was identified with calculating which action would produce the most happiness, either for one person or (and the theory became more vague here) for 'the greatest number' of people. Different versions of the theory substituted 'pleasure' or 'preference' for 'happiness'. Considered as a repudiation of earlier, authoritarian systems of morality, utilitarianism is unexceptionable. And in the sense that it simply advocates rejecting dogma and acting on the 'preferred' theory, the one that has survived rational criticism, every rational person is a utilitarian. But as an attempt to solve the problem we are discussing here, of explaining the meaning of moral judgements, it too has a fatal flaw: *we choose our preferences*. In particular, we *change* our preferences, and we give moral explanations for doing so. Such an explanation cannot be translated into utilitarian terms. Is there an underlying, master-preference that controls preference changes? If so, it could not itself be changed, and utilitarianism would degenerate into the genetic theory of morality discussed above.

What, then, is the relationship of moral values to the particular scientific world-view I am advocating in this book? I can at least argue that there is no fundamental obstacle to formulating one. The problem with all previous 'scientific world-views' was that they had hierarchical explanatory structures. Just as it is impossible, within such a structure, to 'justify' scientific theories as being *true*, so one cannot justify a course of action as being *right* (because then, how would one justify the structure as a whole as being right?). As I have said, each of the four strands has a hierarchical explanatory structure. But the fabric of reality as a whole does not. So explaining moral values as objective attributes of physical processes need not amount to deriving them from anything, even in principle. Just as with abstract mathematical entities, it will be a matter of what they contribute to the explanation – whether physical reality can or cannot be understood without also attributing reality to such values.

In this connection, let me point out that 'emergence' in the standard sense is only one way in which explanations in different strands may be related. So far I have really only considered what might be called *predictive* emergence. For example, we believe that the predictions of the theory of evolution follow logically from the laws of physics, even though proving the connection might be computationally intractable. But the *explanations* in the theory of evolution are not believed to follow from physics at all. However, a non-hierarchical explanatory structure allows for the possibility of explanatory emergence. Suppose, for the sake of argument, that a given moral judgement can be explained as being right in some narrow utilitarian sense. For instance: 'I want it; it harms no one; so it is right.' Now, that judgement might one day be called into question. I might wonder, '*Should* I want it?' Or, 'Am I really right that it harms no one?' – for the issue of whom I judge the action to 'harm' itself depends on moral assumptions. My sitting quietly in a chair in my own home 'harms' everyone on Earth who might benefit from my going out and helping them at that moment; and it 'harms' any number of thieves who would like to steal the chair if only I went elsewhere for a while; and so on. To resolve such issues, I adduce further moral theories involving new explanations of my moral situation. When such an explanation seems satisfactory, I shall use it tentatively to make judgements of right and wrong. But the explanation, though temporarily satisfactory to me, still does not rise above the utilitarian level.

But now suppose that someone forms a general theory about such explanations themselves. Suppose that they introduce a higher-level concept, such as 'human rights', and guess that the introduction of that concept will, for a given class of moral problems like the one I have just described, always generate a new explanation that solves the problem in the utilitarian sense. Suppose, further, that this theory about explanations is itself an explanatory theory. It explains, in terms of some other strand, *why* analysing problems in terms of human rights is 'better' (in the utilitarian sense). For example, it might explain on epistemological grounds why respect for human rights can be expected to promote the growth of knowledge, which is itself a precondition for solving moral problems.

If the explanation seems good, it might be worth adopting such a theory. Furthermore, since utilitarian calculations are impossibly difficult to perform, whereas analysing a situation in terms of human rights is often feasible, it may be worth using a 'human rights' analysis in preference to any specific theory of what the happiness implications of a particular action are. If all this were true, it could be that the concept of 'human rights' is not expressible, even in principle, in terms of 'happiness' – that it is not a utilitarian concept at all. We may call it a moral concept. The connection between the two is through emergent explanation, not emergent prediction.

I am not especially advocating this particular approach; I am merely illustrating the way in which moral values might exist objectively by playing a role in emergent explanations. If this approach did work, then it would explain morality as a sort of 'emergent usefulness'.

In a similar way, 'artistic value' and other aesthetic concepts have always been difficult to explain in objective terms. They too are often explained away as arbitrary features of culture, or in terms of inborn preferences. And again we see that this is not necessarily so. Just as morality is related to usefulness, so artistic value has a less exalted but more objectively definable counterpart, *design*. Again, the value of a design feature is understandable only in the context of a given purpose for the designed object. But we may find that it is possible to improve designs by incorporating a good aesthetic criterion into the design criteria. Such aesthetic criteria would be incalculable from the design criteria; one of their uses would be to improve the design criteria themselves. The relationship would again be one of explanatory emergence. And artistic value, or beauty, would be a sort of *emergent design.*

Tipler's overconfidence in predicting people's motives near the omega point has caused him to underrate an important implication of the omega-point theory for the role of intelligence in the multiverse. It is that intelligence is not only there to control physical events on the largest scale, it is also there to choose what will happen. The ends of the universe are, as Popper said, for us to choose. Indeed, to a large extent the content of future intelligent thoughts *is* what will happen, for in the end the whole of space and its contents will *be* the computer. The universe will in the end consist, literally, of intelligent thought-processes. Somewhere towards the far end of these materialized thoughts lies, perhaps, all physically possible knowledge, expressed in physical patterns.

Moral and aesthetic deliberations are also expressed in those patterns, as are the outcomes of all such deliberations. Indeed, whether or not there is an omega point, wherever there is knowledge in the multiverse (complexity across many universes) there must also be the physical traces of the moral and aesthetic reasoning that determined what sort of problems the knowledge-creating entity chose to solve there. In particular, before any piece of factual knowledge can become similar across a swathe of universes, moral and aesthetic judgements must already have been similar across those universes. It follows that such judgements also contain objective knowledge in the physical, multiverse sense. This justifies the use of epistemological terminology such as 'problem', 'solution', 'reasoning' and 'knowledge' in ethics and aesthetics. Thus, if ethics and aesthetics are at all compatible with the world-view advocated in this book, beauty and rightness must be as objective as scientific or mathematical truth. And they must be created in analogous ways, through conjecture and rational criticism.

So Keats had a point when he said that 'beauty is truth, truth beauty'. They are not the same thing, but they are the same *sort* of thing, they are created in the same way, and they are inseparably related. (But he was of course quite wrong to continue 'that is all ye know on earth, and all ye need to know'.)

In his enthusiasm (in the original sense of the word!), Tipler has neglected part of the Popperian lesson about what the growth of knowledge must look like. If the omega point exists, and if it will be created in the way that Tipler has set out, then the late universe will indeed consist of embodied thoughts of inconceivable wisdom, creativity and sheer numbers. But thought is problem-solving, and problem-solving means rival conjectures, errors, criticism, refutation and backtracking. Admittedly, *in the limit* (which no one experiences), at the instant when the universe ends, everything that is comprehensible may have been understood. But at every finite point our descendants' knowledge will be riddled with errors. Their knowledge will be greater, deeper and broader than we can imagine, but they will make mistakes on a correspondingly titanic scale too.

Like us, they will never know certainty or physical security, for their survival, like ours, will depend on their creating a continuous stream of new knowledge. If ever they fail, even once, to discover a way to increase their computing speed and memory capacity within the period available to them, as determined by inexorable physical law, the sky will fall in on them and they will die. Their culture will presumably be peaceful and benevolent beyond our wildest dreams, yet it will not be tranquil. It will be embarked upon the solution of tremendous problems and will be split by passionate controversies. For this reason it seems unlikely that it could usefully be regarded as a 'person'. Rather, it will be a vast number of people interacting at many levels and in many different ways, but *disagreeing*. They will not speak with one voice, any more than present-day scientists at a research seminar speak with one voice. Even when, by chance, they do happen to agree, they will often be mistaken, and many of their mistakes will remain uncorrected for arbitrarily long periods (subjectively). Nor will the culture ever become *morally* homogeneous, for the same reason. Nothing will be sacred (another difference, surely, from conventional religion!), and people will continually be questioning assumptions that other people consider to be fundamental moral truths. Of course, morality, being real, is comprehensible by the methods of reason, and so every particular controversy will be resolved. But it will be replaced by further, even more exciting and fundamental controversies. Such a discordant yet progressive collection of overlapping communities is very different from the God in whom religious people believe. But it, or rather some subculture within it, is what will be resurrecting us if Tipler is right.

In view of all the unifying ideas that I have discussed, such as quantum computation, evolutionary epistemology, and the multiverse conceptions of knowledge, free will and time, it seems clear to me that the present trend in our overall understanding of reality is just as I, as a child, hoped it would be. Our knowledge is becoming both broader and deeper, and, as I put it in Chapter 1, depth is winning. But I have claimed more than that in this book. I have been advocating a particular unified world-view based on the four strands: the quantum physics of the multiverse, Popperian epistemology, the Darwin–Dawkins theory of evolution and a strengthened version of Turing's theory of universal computation. It seems to me that at the current state of our scientific knowledge, this is the 'natural' view to hold. It is the conservative view, the one that does not propose any startling change in our best fundamental explanations. Therefore it ought to be the prevailing view, the one against which proposed innovations are judged. That is the role I am advocating for it. I am not hoping to create a new orthodoxy; far from it. As I have said, I think it is time to move on. But we can move to better theories only if we take our best existing theories seriously, as explanations of the world.

Bibliography

EVERYONE SHOULD READ THESE

Richard Dawkins, *The Selfish Gene*, Oxford University Press, 1976. [Revised edition 1989.]

Richard Dawkins, *The Blind Watchmaker*, Longman, 1986, Norton, 1987; Penguin Books, 1990.

David Deutsch, 'Comment on "The Many Minds Interpretation of Quantum Mechanics" by Michael Lockwood', *British Journal for the Philosophy of Science*, 1996, Vol. 47, No. 2, p. 222.

David Deutsch and Michael Lockwood, 'The Quantum Physics of Time Travel', *Scientific American*, March 1994, p. 68.

Douglas R. Hofstadter, *Gödel, Escher, Bach, an Eternal Golden Braid*, Harvester, 1979, Vintage Books, 1980.

James P. Hogan, *The Proteus Operation*, Baen Books, 1986, Century Publishing, 1986. [Fiction!]

Bryan Magee, *Popper*, Fontana, 1973, Viking Penguin, 1995.

Karl Popper, *Conjectures and Refutations*, Routledge, 1963, HarperCollins, 1995.

Karl Popper, *The Myth of the Framework*, Routledge, 1992.

FURTHER READING

John Barrow and Frank Tipler, *The Anthropic Cosmological Principle*, Clarendon Press, 1986.

Charles H. Bennett, Gilles Brassard and Artur K. Ekert, 'Quantum Cryptography', *Scientific American*, October 1992.

Jacob Bronowski, *The Ascent of Man*, BBC Publications, 1981, Little Brown, 1976.

Julian Brown, 'A Quantum Revolution for Computing', *New Scientist*, 24 September 1994.

Paul Davies and Julian Brown, *The Ghost in the Atom*, Cambridge University Press, 1986.

Richard Dawkins, *The Extended Phenotype*, Oxford University Press, 1982.

Daniel C. Dennett, *Darwin's Dangerous Idea: Evolution and the Meanings of Life*, Allen Lane, 1995; Penguin Books, 1996.

Bryce S. DeWitt and Neill Graham (eds), *The Many-Worlds Interpretation of Quantum Mechanics*, Princeton University Press, 1973.

Artur K. Ekert, 'Quantum Keys for Keeping Secrets', *New Scientist*, 16 January 1993.

Freedom and Rationality: Essays in Honour of John Watkins, Kluwer, 1989.

Ludovico Geymonat, *Galileo Galilei: A Biography and Inquiry into his Philosophy of Science*, McGraw-Hill, 1965.

Thomas Kuhn, *The Structure of Scientific Revolutions*, University of Chicago Press, 1971.

Imre Lakatos and Alan Musgrave (eds), *Criticism and the Growth of Knowledge*, Cambridge University Press, 1979.

Seth Lloyd, 'Quantum-mechanical Computers', *Scientific American*, October 1995.

Michael Lockwood, *Mind, Brain and the Quantum*, Basil Blackwell, 1989.

Michael Lockwood, 'The Many Minds Interpretation of Quantum Mechanics', *British Journal for the Philosophy of Science*, 1996, Vol. 47, No. 2.

David Miller (ed), *A Pocket Popper*, Fontana, 1983.

David Miller, *Critical Rationalism: A Restatement and Defense*, Open Court, 1994.

Ernst Nagel and James R. Newman, *Gödel's Proof*, Routledge 1976.

Anthony O'Hear, *Introduction to the Philosophy of Science*, Oxford University Press, 1991.

Roger Penrose, *The Emperor's New Mind: Concerning Computers, Minds, and the Laws of Physics*, Oxford University Press, 1989.

Karl Popper, *Objective Knowledge: An Evolutionary Approach*, Clarendon Press, 1972.

Randolph Quirk, Sidney Greenbaum, Geoffrey Leech and Jan Svartvik, *A Comprehensive Grammar of the English Language*, 7th edn, Longman, 1989.

Dennis Sciama, *The Unity of the Universe*, Faber & Faber, 1967.

Ian Stewart, *Does God Play Dice? The Mathematics of Chaos*, Basil Blackwell, 1989; Penguin Books, 1990.

L. J. Stockmeyer and A. K. Chandra, 'Intrinsically Difficult Problems', *Scientific American*, May 1979.

Frank Tipler, *The Physics of Immortality*, Doubleday, 1995.

Alan Turing, 'Computing Machinery and Intelligence', *Mind*, October 1950. [Reprinted in *The Mind's I*, edited by Douglas Hofstadter and Daniel C. Dennett, Harvester, 1981.]

Steven Weinberg, *Gravitation and Cosmology*, John Wiley, 1972.

Steven Weinberg, *The First Three Minutes*, Basic Books, 1977.

Steven Weinberg, *Dreams of a Final Theory*, Vintage, 1993, Random, 1994.

John Archibald Wheeler, *A Journey into Gravity and Spacetime*, Scientific American Library, 1990.

Lewis Wolpert, *The Unnatural Nature of Science*, Faber & Faber, 1992, HUP, 1993.

Benjamin Woolley, *Virtual Worlds*, Basil Blackwell, 1992; Penguin Books, 1993.

Index

Numbers in **bold** refer to principal or defining occurrences; those in *italics* refer to figures.